Is it Hot in Here?

Is it Hot in Here?

✦

The simple truth about global warming

Nathan Todd Cool

iUniverse, Inc.

New York Lincoln Shanghai

Is it Hot in Here?
The simple truth about global warming

Copyright © 2006 by Nathan Todd Cool

iUniverse books may be ordered through booksellers or by contacting:

iUniverse
2021 Pine Lake Road, Suite 100
Lincoln, NE 68512
www.iuniverse.com
1-800-Authors (1-800-288-4677)

ISBN-13: 978-0-595-40622-7 (pbk)
ISBN-13: 978-0-595-84989-5 (ebk)
ISBN-10: 0-595-40622-X (pbk)
ISBN-10: 0-595-84989-X (ebk)

Printed in the United States of America

To everyone who's ever pondered our planet's climate change predicament, scratched their head and wondered just what in the heck is really going on.

Humankind has not woven the web of life. We are but one thread within it. Whatever we do to the web, we do to ourselves. All things are bound together. All things connect.

—Chief Seattle
Suquamish and Duwamish
Native American tribes
1786-1866

Contents

Acknowledgements

o o o o o o o o o o o o o o o o o o o o o o o o o o o o o o o o o o o

Sometimes our light goes out but is blown into flame by another human being. Each of us owes deepest thanks to those who have rekindled this light.
—Albert Schweitzer

No book can be entirely credited to an author alone, as so many people assist in making the pages possible. While sitting here writing these words of thanks, reminded by a few aches and pains in my writer's cramped wrists and hands, telling me I've persevered a plentitude of pages, plethora of paragraphs, scores of sentences and oodles of words, I feel humbled knowing that this book was made possible by more than just my typing-fatigued fingers, and nearly a year's worth of research. I owe a debt of gratitude to many others.

I have many people to thank for assisting and motivating me throughout the past year while working on *Is it Hot in Here?*, including those who regularly read my reports and have voiced their opinions. Such was the impetus for this book. To all my readers, thank you for your continued support.

A big thanks goes out to Dr. David Field for his valuable input on decadal oscillation research on seafloor sediment sampling and manuscript review. Thank you David for setting aside so much of your valuable time, especially during your research in Peru, to patiently read through my work, and for pursuing the search for answers of our hidden, climatic past.

I'd like to extend a special thanks to Dr. William M. Connolley, Senior Scientific Officer and Climate Modeler from the British Antarctic Survey for his review of my work and valuable input, especially in regards to Vostok, water vapor and GHG's, and melting sea ice.

I'd like to extend my gratitude to Michael Larsen for his input on this book's title, and encouragement to continue my work on this book, and see it through to the end.

My brother Dan Cool deserves more than mere thanks for his megabytes of assistance in countless backups of this book's manuscript. Without his help, I don't know if I would have slept as well as I have over the past year.

To say I'm grateful for the review by Josh Rubenstein of KCBS/KCAL, especially while suffering a nasty cold and filling the shoes of two meteorologists at the same time would be an understatement. Josh, all I can say is thank you, thank you, thank you, thank you!

I'd also like to thank those who took time out of their busy schedules to review this book's manuscript during its infancy, and for looking beyond the original typos and other grammatical errors riddling the first few drafts and providing me their reviews. A special, heartfelt thanks goes out to Aaron Kenedi from *Shift Magazine* as well as Tom Turner from *Earthjustice*.

And last, but certainly not least, I'd like to extend a special thanks to Gwen Mickelson from the *Santa Cruz Sentinel* for not only being such a patient and kind interviewer, but also for—with such short notice—taking the time to peruse these pages and provide me her review. Thank you Gwen, and forever may you ride.

Introduction

Everybody talks about the weather, but nobody does anything about it.
—*Mark Twain*

In 2005, residents of Nikolski were stunned to see a hummingbird visit their Aleutian Chain island near the Alaska Peninsula. Inhabitants on the South Pacific island of Tegua and the Carteret Islands off Papua New Guinea were forced to higher ground after losing a battle with rising sea levels that washed away their homes and tainted their drinking water. Seals native to California were seen in 2005 in the far reaches of the Northern Pacific. New species of insect-borne parasites killed seventy reindeer in Norway around this same time as well. And who'd have thought that Eskimos would need sunscreen, but as global temperatures rise, so have the occurrences of skin cancer among Canadian Inuits.[1]

Sounds like global warming, right? Well, maybe yes, maybe no. It all depends on whom you ask. The topic of global warming is about as controversial as animal rights, genetic engineering, gun control, the death penalty, racial profiling, evolution, or The Patriot Act. Any topic that touches on theories or ideologies will inevitably be accompanied by a diversity of opinions, making it a tough job for anyone trying to dig through the myriad material, stats, data, and news on the subject to find unbiased facts on their truth-seeking sojourn. I encountered such a long and winding road on a recent voyage to the land of answers on the topic of global warming—it was a long, strange trip indeed.

Recently, I received a lot of feedback from some articles I wrote on the topic of climate change, and how global warming may or may not be affecting our planet—in particular, how hurricane seasons may be affected, and the resultant impacts on beach, surf, and coastal communities. After enduring a record-breaking hurricane season in 2005, and being deluged with increasingly more articles popping up in the media regarding the global warming debate, I went in search of answers and published my findings. It was a somewhat risky move, kind of like bringing up religion or politics in mixed company; you're sure to get some flack with your kudos.

Although for the most part I chose to seat myself firmly on the fence and not hop off into either the "sky-is-falling camp" or the opposing "everything-is-dandy gang," I found compelling arguments coming from both sides. Subsequently, I received numerous emails from a range of readers; those who felt global warming is real, those who felt it's just a lot of hype, and from those who are curiously seeking to make sense out of this hotly debated topic. I could tell from the responses I received that many readers were in the same boat of confusion that I was on my journey across the disorienting sea of information searching for answers on global warming. Today, there's an overabundance of information out there on global warming, but for every bit of evidence that seems to point to a world that's warming due to man's hand, persuasive arguments are made to discount such data.

Over the years, science has had a rough go at things as a tenacious few in search of answers fought uphill battles in a world saturated with skepticism to prove their points. Who would believe that the world is round? Didn't the Sun and planets revolve around the Earth? Who'd have thought that man could fly, or land on the Moon? Even Thomas Jefferson, when discussing the possibility that meteorites could fall to Earth commented, "I would more easily believe that two Yankee professors would lie, than that stones would fall from heaven."[2] Science has come a long way over the course of humankind's existence on our planet. Nevertheless, history shows us that politics, religion and just about any other subject holding credos or agendas will find advocates fighting tooth and nail to justify their views. In so doing, defenders of a faith tend to overlook some of the facts. In the global warming debate, this goes for both sides.

For the most part, reader responses regarding my recent articles were positive, giving me kudos for staying on that proverbial fence, and not hopping off onto a soapbox to tout my beliefs over the subject. Yet a few of the reader responses accused me of being a tree-hugger, sympathetic to Democrats, who mostly seem to sit on the pro global warming side of the arena, and feel that "the other side" is in bed with "big oil." A few other readers thought of me as another right-winger who won't face the fact that we're all doomed, and big oil, in cahoots with the U.S. Government, is tearing our planet apart. This variety of viewpoints isn't much different from any other contentious topic where two overlying forces form an opposing dichotomy. The Scopes Monkey trial of 1925—where the battle of science vs. religion was presented to a jury—and even recent laws over stem cell research have been split along the dividing lines separating bipartisan views and fundamental belief systems. Global warming, in many respects, has befallen this same fate as well.

Being a man who bolsters his beliefs on science and not politics, I don't subscribe to the "my team" mentality. I'm not a sports fan, and I don't root for any one team. In fact, I'd have to admit, I can be a real wet blanket at a Super Bowl party. When one team scores a touchdown, I can't help but wonder what the other team did wrong, and question what they could have done to avert a future loss. I usually keep my mouth shut during the game though, as I enjoy being invited back to gorge on chips, been dip, and hot wings amongst the company of my friends. Nevertheless, my inclination to not be inclined, and to look at both teams as having equal merits, goes for my political beliefs as well. I don't make my decisions based on how a particular political party may think, or what they promote. Call me a rebel if you will, but I like to form my own opinions. After all, since we live in a free country, freedom of belief is our inalienable right. Still, at some point, I will have to step down from this uncomfortable fence I'm sitting on, and actually commit to a decision: Is global warming real? Or is this just a lot of hooey?

Feeding the fuel to rationalize my fear of commitment is a profusion of confusion surrounding the heated debate over global warming. If there were no confusion on the matter, then we'd pretty much have a consensus, and we'd all get along, agreeing one way or the other. But in a world where media is streamed into our homes as it happens, and a plethora of ratings-hungry news agencies vie for our attention, we're bombarded by influential arguments coming in from both sides of the global warming trenches. For instance, instead of composing the gloom-and-doom apocalyptic opening paragraph for this chapter, I could have written something like the following, which is—technically speaking—just as true:

> Island residents in the South Pacific fled to higher ground in 2005 as scientists blame sinking atolls for disappearing islands. Germany and Siberia now face the coldest winters on record in 2006. Scientists, who are finding no significant increase in man-made CO_2, are also finding that not only are we entering an oceanic decadal oscillation, but also a millennial oscillation that can be blamed for recent changes in our weather.[3]

Personally, I prefer my original opening paragraph, so I left it as is. It has a higher "shock and awe" value and a nice ring to it as well. The media, playing a similar game, provide us with conflicting and oftentimes confusing information on global warming, making it difficult at times to discern the truth—whatever that may be. In our world of action-packed, ratings hungry, sensationalistic story spinners, opinions can lead to bias, and oftentimes not science.

In late 2004, the organization for Fairness and Accuracy in Reporting (FAIR) published a report titled *Journalistic Balance as Global Warming Bias: Creating controversy where science finds consensus*. In this report, FAIR, through their independent study, found that the U.S. media coverage of global warming often becomes a form of informational bias with a lot of "he said/she said" reporting by believers and skeptics alike. In FAIR's analysis of over 3,000 articles selected from the New York Times, Washington Post, Los Angeles Times and Wall Street Journal covering the topic of global warming, they found that only about half of the articles gave equal attention to both views that: (a) humans may contribute to global warming and (b) that climate change could be exclusively the result of natural fluctuations. The remaining articles however, the other nearly 50%, were swayed to one side or the other, showing bias.

It probably comes as no surprise that the media could put a spin on a story, and that different news organizations may have dissimilar views on any given subject. Election year campaigns are proof enough of that. What is surprising however, is how articles are doing this in the war of words on global warming. For example, an article from the Washington Post in 1995 previewed the First Conference of the Parties (COP1) to the U.N. Framework Convention on Climate Change in Berlin.[4] The article described *"the lack of international consensus on the causes and hazards of global warming"* before turning to the concerns of the residents of the Maldive Islands, a low-lying country that could be submerged if rising tides from global warming continue. After citing the distress of the Maldive president—whose land is slowly being subsumed by the sea—the article closes by saying:

> On the other hand, some skeptical meteorologists and analysts assert that global warming reflects a natural cycle of temperature fluctuation and cannot be decisively tied to human actions. "As far as we are concerned, there's no evidence for global warming, and by the year 2000 the man-made greenhouse theory will probably be regarded as the biggest scientific gaffe of the century," Piers Corbyn, an astrophysicist at London's Weather Action forecasting organization, told the Reuters news agency.

Although the Washington Post article did show both sides of the coin, it seemed to bolster the views of global warming skeptics with its closing statements. And then the tables turned in another article published in the same periodical (the Washington Post) in January 2006, which was titled *Study: Global Warming May Raise Sea Levels*. This article, originating from the Associated Press said[5]:

> Most scientists believe greenhouse gases from human activities like coal-burning power plants cause global warming by trapping solar heat in the atmosphere.

And later on, the article states:

> Many island nations are already feeling the impact of rising seas. In Vanuatu and Papua New Guinea in the Southwest Pacific, increased sea levels have forced hundreds of islanders to abandon vulnerable coastal homes for higher ground, according to the United Nations and news reports.

Perhaps the editors at the Washington Post have changed their views in the past few years, or maybe they are trying to report in a fair and balanced manner. Recently though, another story came into the news that brought on even more uncertainty. An article published by National Geographic News[6] titled *Frog Extinctions Linked to Global Warming*, talked about a study being published in the journal *Nature*, that seems to show a direct link to global warming and the decline in harlequin frog populations in Central and South America. But when turning to Fox News[7], an article published on January 24, 2006 seemed to dispute this with a story titled *Study: Global Warming Not Killing Off Arizona Frogs*. Granted, Arizona is not in the same neck of the woods as the other study on declining frog populations, but the vast difference between these two reports makes one wonder if global warming can really kill off any amphibian species at all. In both articles, a disease-causing chytrid fungus was the alleged perpetrator. Interestingly enough however, even though the *National Geographic News* and *Nature* articles that linked global warming to chytrid-induced frog deaths were published twelve days before the Fox News story came out, the story on Fox didn't mention these other articles.

I can empathize with any investigative reporter. Gathering all available data to produce a story can be a laborious task to say the least. Nevertheless, the global warming-linked frog report published in *Nature* was broadcast throughout the media when it became available, and it would have been difficult to miss it. In fact, doing a search using Google News[8] with the key words "frogs global warming" revealed 233 different stories covering this topic, most of which were related to the recently released global warming-linked report in *Nature*. Still, giving Fox News the benefit of the doubt, they may have merely overlooked this information. And with all fairness to the media in general, of all things to report on, theories, especially those involving long-range climate effects, can be tricky.

As Chief Forecaster for the WaveCast® service at WetSand.com, I, for the most part, have a fairly easy job in the prediction business. All I need to do is track storm systems forming in the oceans around the globe, and by working through some equations, I can formulate, within a fairly high degree of accuracy what surfing conditions will be like many days in advance for just about any spot on the planet. I'm not some kind of shaman or magician; it's all a matter of mathematics, and dealing with known laws of physics. Anyone can do it, and I even wrote a book[9] showing how anyone with access to the Internet and a calculator can forecast wave arrival times, size, and other various factors that pique the interests of board riders. When it comes to tracking surf conditions one to two weeks in advance, all that anyone has to do is check in with Mother Nature to see what she's up to, and then let the laws of science do the rest. That's the easy part. One of my more difficult responsibilities though, is to gather long-range data to compile predictions on how ocean climates may affect local weather, surf and beach conditions for upcoming seasons. This is a bit thornier than just plugging in some numbers and calculating out simple principles of fluid dynamics. When it comes to long-range seasonal forecasts, science often mixes with assumptions, just like it does in the global warming debate.

Once every quarter, I pay a visit to a radio talk show hosted by Marc Kent called Let's Talk Surfing where Marc and I discuss what lies ahead in our oceans' cloudy crystal ball. This has been a particularly popular spot on Marc's show since the astounding El Niño of 1997-1998. During that winter season, I was overwhelmed tracking numerous, large-scale storm systems that formed in the Western Pacific and grew to astronomical magnitudes as they traversed eastward across the Pacific at lower than normal latitudes, resulting in surf reaching 50 feet along northern shores of Hawaii and Northern California. This and the media frenzy concerning El Niño around that time caught the attention of more than just those wanting to paddle out and hang ten on "the big one." That season seemed to mark a change in how many perceived our planet's climate, and almost immediately, a consensus was formed to attribute this seasonal phenomenon to global warming. But being the man of science that I am, and not one to easily fall victim to mob or media-hype mentalities, I felt the verdict was still out, and that more proof was needed.

As I waited, a few years later, El Niño's antithesis came to fruition: a La Niña. Waters turned from tepid to frigid. My niece April from Pennsylvania came to visit her California uncle around this time, and we both found the beach to be a bit too chilly in August, and the water far to brisk for even the quickest of dips. My niece's California experience was not at all what she expected, and she quickly understood why her uncle donned a wetsuit during those summer months. But more importantly, when looking for evidence to tie El Niño to global warming, this cold-water episode just didn't pass muster, or at least so it seemed.

What I did learn from all of this (and more) is that it pays to be vigilant and weigh all sides of an issue before making a definitive claim that something is indeed true. Had I announced in 1998 that global warming was the cause for the intense weather and pounding surf during that El Niño winter, surfers a few years later sitting in chilly waters with chattering teeth would have thought I was off my rocker. Interestingly enough though, even that intense cold-water episode could be pointing to climate change. When it comes to global warming, it's oftentimes one of those, "damned if you do, and damned if you don't" kind of things, making this topic an even muddier mess to fish through to find answers.

While doing my darndest to be a man of science over the years, I've tried hard to keep an open mind and not pass judgment too quickly. Theories often change over time. Even the theory of evolution has been, well, evolving, and is yet to be completely solidified. New human species have been found since the "missing link" hypothesis of yore, and some sectors of religion are conceding that the world may indeed be older than 5,000 years, and that possibly Intelligent Design could be afoot. As I write these words, a tiny space probe is on its way to Pluto to see if this supposedly ninth world in our solar system really is a planet, and not a bulky, ice-crusted asteroid from the Kuiper Belt, as many astronomers now believe. As time marches on, science progresses and we learn more about the world and universe in which we live. Global warming, with all of its apparently damning evidence, is still in the early, investigative stages, and needs more time to become a science that is fully understood. Nevertheless, much of the evidence of global warming is indeed persuasive. Still, with open minds, and being able to analyze the facts, each of us can exercise our inalienable rights of freedom of belief to form our own opinions—based on facts, and not "my team" mentalities, hunches, or superstitions.

To help in this matter, I've constructed this book as not just a primer on global warming, and not a boring volume of scientific reference either; instead, I've set out to explain, in easy to understand terms, the science behind global warming, and the facts known about it thus far. I'll also be showing data and opinions from both sides of the global warming debate, in addition to why each argument seems convincing.

Chapter 1 will start out with a brief tutorial on global warming—what is it, who first noticed it, and other such tidbits. Subsequent chapters will discuss what effects are (or are not) being felt right now from climate change, and how this is being interpreted in various political and scientific sectors. I'll also discuss both sides of the global warming debate and drill into the sciences surrounding the statements touted from trenches on both sides of the battlefield. Then I'd like to talk about what may happen in the future, and what steps we can take—and are currently being pursued—to prepare or possibly avert a global warming catastrophe, should one occur.

As this book will point out, there are many uncertainties surrounding global warming, yet some things are well assured. With an open mind, I'll be covering the ins and outs of what we do and do not know about global warming in hopes that we can answer that one, single burning question: Is it hot in here?

1

Climate Change 101

The most exciting phrase to hear in science, the one that heralds new discoveries, is not Eureka! (I found it!) but rather, "hmm.... that's funny..."
—*Isaac Asimov*

In a day and age of political correctness, it's often more acceptable to refer to the warming of our planet as "global climate change." Just as a policeman or policewoman is now a "police officer," airline stewards and stewardesses are called "flight attendants," impotence is referred to as "erectile dysfunction," Merry Christmas is uttered as "Happy Holidays," illegal aliens are booked as "undocumented immigrants," and Star Trek fans now "boldly go where no *one* has gone before," the term "global warming" seems offensive to some folks. Thanks in large part to Hollywood sensationalism, hearing the words *global warming* stirs doomsday images from such movies as *The Day After Tomorrow* where, as though overnight, the world will be hit by a plague of nasty weather with prolific hurricanes, tornadoes, tidal waves, and the beginning of a new ice age. Others cringe at the term global warming, envisioning ice caps melting so rapidly that they force all of humankind to run in panic to flee rising seas while flesh is burned off our bones from the ever-increasing heat of the Sun.

Calling it *climate change* does give this phenomenon of planetary heating a bit more of a (excuse the pun) *warm* and fuzzy appeal, and I have to admit, it does have a less apocalyptic tone to it. Nevertheless, the term *global climate change* can allude to any phenomenon that would change our weather, and doesn't necessarily describe the possibility that our world could be warming. So, at the risk of being politically incorrect, throughout this book I will often refer to this phenomenon as global warming and use this term synonymously with climate change.

Setting all monikers, sobriquets, handles and labels aside though, just what exactly *is* this thing called global warming, who first noticed it, and why is it getting so much darn attention these days?

What is Global Warming?

Global warming is something that's been under the scientific community's proverbial microscope for quite a long time, with its most recognized genesis of discovery occurring in the 19th century. Back in those days, long before most people were aware of global weather, two primary topics regarding the climate came of interest: weather related to the logging industry, and ice ages.

Many Americans thought that cutting away forests for lumber brought about more rain to affected, deforested regions. Although this may have just been a matter of witnessing more mudslides from these now treeless areas, having no footing to hold back the soil when the rains did eventually come, the logging industry was seen by many as a business sector harming the environment. Big business was being linked in the minds of many as a great evil against the Earth, turning once picturesque landscapes into unsightly blights that were further decimated by rivers of mud sliding down now barren hillsides. Although this concern may not have been directly linked to any weather phenomena, it did stir the curiosity of many, and from curiosity comes a search for answers. With more people tuning into the environment in the early 19th century, filled with curiosity about weather and effect causality, scientific research gained momentum.

For the most part, scientists in the 19th century who were researching long-term weather effects were not really interested in a warming planet. In fact, they were looking in the opposite direction: ice ages. If there was one thing that could possibly impair the world in which they lived, it was thought that the hand of man—or natural causes—could bring about another nasty cold spell. Research in the 19th century was showing, without a doubt, that our planet does indeed go through temperature fluctuations from time to time. This research led most scientists to realize that the skies above us were being affected by something, which in turn was changing Earth's climate. What they found was intriguing.

Over time, tenacious researchers were able to identify that mysterious something that was changing the world's climate. The key suspect, which is still being studied today as the primary contributor for global warming is what is known now as "atmospheric greenhouse gases," or just "greenhouse gases" for short. These gases, which I'll describe in more detail later on (in *When Termites Fart*), include natural occurring elements such as water vapor, carbon dioxide, and other lighter-than-air substances. But these greenhouse gases get a bad rap. In fact, if it weren't for greenhouse gases in our atmosphere, we wouldn't be here today. You see, the Sun does a good job of heating our planet, but the heat it generates would be quickly swept away from Earth if it were not for the greenhouse gases, which trap heat rising up from our planet, making sure it doesn't all escape. These gases act like the glass panes in a greenhouse that hold in heat; hence, the name: greenhouse gases.

Luckily, our planet was made with a nice, light coating of a greenhouse gas-rich atmosphere, which Mother Nature uses to wrap us up like a warm, cozy blanket. This makes life possible here on our tiny, blue, and otherwise chilly planet. Other planets have greenhouse gases, but Earth has just the right amount to keep us happy. If men were from Mars and women from Venus—two other planets that have greenhouse gases in their atmospheres—things would be quite different, even more so than they already are here on Earth. The male population on Mars—where greenhouse gases are nearly nonexistent—would have to endure air temperatures that reach a frigid -200°F, but will at times climb to the balmy Martian 32°F on a sunny day at just the right spot. The ladies, living on Venus—where greenhouse gases are extremely thick—would have no complaints about needing an extra blanket or turning up the thermostat to keep warm, as they would be living in temperatures running around 850°F, on a cool day. Although Mars and Venus have greenhouse gases in their atmospheres, their concentrations are much different than we have here on Earth, and is why we never have to worry about alien invasions from Mars or Venus—no one lives there, no one can.

With the atmosphere carefully balanced with greenhouse gases here on the third blue rock from the Sun, we maintain an even keel on the thermometer. Nature itself creates some of these greenhouse gases, and nature can do a pretty good job at keeping them in check, recycling these gases so that we maintain a balance in our atmosphere. Nature's recycling process keeps our atmospheric blanket at just the right thickness and density. But Mother Nature's airy blanket is actually quite delicate, and relatively speaking, thin. Our atmosphere extends upward for about 120 miles, which may seem reasonably bounteous sitting here on terra firma, but to put this in perspective, if you shrank our planet down to the size of the standard, everyday desktop globe, our entire atmosphere would be no thicker than just a couple coats of varnish.[10] If this diaphanous blanket of sky above us becomes too heavily laden with greenhouse gases, so much so that the gases can't be recycled by the forces of nature, then the gases will build up, and we're in for trouble.

Since the beginning of the Industrial Revolution, greenhouse gases have been on the rise, and nature hasn't been able to keep up on its recycling efforts. More and more greenhouse gases are being fed into the atmosphere, and they're accumulating. This is alarming news to many in the scientific community, fearing that this thicker atmospheric blanket will trap far too much heat. Many adverse things could come as a result of this, which I'll discuss in greater detail throughout the book. In a nutshell though, increasing greenhouse gases could in turn increase global temperatures, which could lead to rising sea levels and significant changes in the Earth's weather. Global warming could alter our Earth's forests, crop yields and water supplies as well. It could also affect the health of humans, animals, and myriad ecosystems. And a warming world could have long-lasting, detrimental impacts on world economies and the wellbeing of billions of people around the globe for centuries.

Although *this* may sound like fantastic Hollywood movie sensationalism, global warming's *potential* for harm is real, and alarming, which is why it has gained so much attention in the scientific community, political arena, and most of the world for that matter. But what exactly is to blame for the increase in greenhouse gases? And is this increase something that will indeed harm our environment and way of life?

Many believe that humankind is to blame for the increase in greenhouse gases, while others show evidence to the contrary and dispute that this is merely a natural cycle of our planet, and equilibrium will eventual come. After all, our planet has gone through ice ages, so we know that our world's weather does change from time to time. But back in the day of wooly mammoths and saber-toothed tigers, there weren't nearly as many humans on the planet, and those that did populate our Earth didn't have the greenhouse gas-emitting volume that we possess today. This is certainly a conundrum, and figuring it all out has not been easy. Many questions have been spawned from the global warming quandary, and most often, these questions are not at all too easy to explain. Many who even do agree that our planet is warming due to increased greenhouse gases, still differ in their opinions on just what outcomes we can expect.

Scientists are a skeptical bunch; that's what makes science, science. No one single discovery can be taken as fact. Scientists like evidence, and lots of it. To establish proof or at least theories of global warming has been an especially arduous uphill battle throughout history. What's been found over time though has been startling.

DISCOVERY

In 1798, on the northern shores of Egypt in the city of Alexandria stood one of history's most infamous generals, Napoleon Bonaparte. Bloody battles would soon commence on both land and sea with the intention to seize Egypt in the name of France. But this voyage was not only for the intent of war; it was for science as well. Soldiers and officers were not the only ones accompanying Napoleon on this expedition; a young man, a son of a tailor who would later indulge in scientific explorations that would serve as impetus to climate change research, accompanied the stout French general. This man was Joseph Fourier.

Fourier would later be made governor of Lower Egypt, and while being embedded deep within the old Ottoman Empire, cut off from France by the English fleet battling Napoleon's navy, Fourier kept busy organizing workshops, which the French army relied on for their munitions of war. Around this same time, Fourier also contributed several mathematical papers to the Egyptian Institute, which Napoleon founded at Cairo. And then in 1824, Joseph Fourier, while poking around in thermodynamics, discovered something unique about our atmosphere.

Fourier was conducting numerous experiments on heat and in particular, its effect on the surface temperature of the Earth. In an essay written in 1824, Fourier showed how he established a new theory called "planetary energy imbalance." In this theory, Fourier explained that planets obtain energy from a number of sources that cause temperatures to increase, but planets also lose energy by something that Fourier called "dark heat." With this combination of energy being directed at a planet (e.g. sunlight shining on the Earth), and some, but not all of this energy escaping, a temperature balance on that planet is reached. Fourier pointed out that a planet's atmosphere, as it thickens, shifts the balance toward higher temperatures.

Now this may seem like common sense to anyone who remembers their grade school science classes. But we're talking about 1824, and things were a lot different back in those days. Morse wouldn't be demonstrating his electric telegraph for another twenty years, and Alexander Graham Bell wouldn't even be born for another twenty-three years. John Quincy Adams was a year away from being elected as the sixth president of the United States to succeed Andrew Jackson. And only one year before Fourier's discovery, William Sturgeon invented the world's first electromagnet.

Scientists were intrigued by Fourier's work back in the early 1800s, and tossed around his theories and experiments for many years. Many came to accept his theories, including one man who was not just a lover of science, but of nature as well: John Tyndall.

We often envision Scientists as somber folk, sporting white lab coats while surrounded by beakers and test tubes. This wasn't the case though with John Tyndall. Tyndall loved the outdoors and was an ardent Alpinist, giving him up-close and personal encounters with glaciers on many occasions. In fact, in 1860, he made the first assent of the Weisshorn, a towering mountain in the Swiss Alps with an incredible, nosebleed altitude of 14,780 feet. Seeing glaciers first hand, Tyndall, familiar with the ice age research conducted within his era, was curious as to what could bring about such massive melting as to leave only glaciers on the highest of mountain peaks where only tens of thousands of years prior, ice had covered nearly all of Northern Europe in the grips of an ice age.

Tyndall, inspired by ice age theories, melting glaciers, and prior work by colleagues such as Joseph Fourier, began to study the radiation properties of various gases around 1859. Tyndall was able to conclude that oxygen, nitrogen, and hydrogen were almost transparent to heat, while other gases seem to block it. One of the toughest radiation impeding gases was what used to be called "carbonic acid," which is now known as carbon dioxide.

Then in 1896, a chemist from Sweden, Svante August Arrhenius, made a significant breakthrough, partially based on Fourier's original findings, and the research conducted by contemporaries like John Tyndall.

Arrhenius, a Nobel Prize laureate, was a mathematical genius. He is well known to many in the scientific community as the first person to propose a remarkable equation to formulate temperature dependencies on chemical reaction rates, which is now known as the "Arrhenius Equation." But what Svante Arrhenius is most likely known for is his theories to explain ice ages.

Arrhenius was able to put many pieces of the puzzle together, and came up with the idea that changes in the levels of carbon dioxide in the atmosphere could substantially alter the Earth's surface temperature through a kind of greenhouse effect. He was influenced by the work conducted by Joseph Fourier, and used infrared observations of the moon gathered from the Allegheny Observatory in Pittsburgh to calculate the absorption of carbon dioxide and water vapor. Using a new principle called "Stefan's Law" Arrhenius formulated his "greenhouse law," which was often coined "the hothouse theory."

Arrhenius later published many books on the subject including *Worlds in the Making* in 1908, where he suggested that the human emission of carbon dioxide could be strong enough to prevent the world from entering a new ice age, and that a warmer Earth would be needed to feed the rapidly increasing population. Arrhenius didn't seem to be concerned that the planet would overheat; instead, his trepidation was that Earth would need to keep heating up so that we all wouldn't freeze to death in what may be an inevitable glacial epoch. Since this seemed like a more viable concern around the turn of the 20th century than that of an overheated planet, his hothouse theory gained a lot of attention, not just in the scientific community, but among many laypeople as well. In fact, another respected scientist in the late 1800s, Walter Nernst, proposed the bright idea of setting fire to useless coal seams in order to release enough carbon dioxide into the atmosphere, and deliberately warm the Earth's climate. Luckily, for us, Nernst's idea never came to fruition.

Nevertheless, until the early 1960s, most scientists dismissed Arrhenius' hot-house theory as implausible. The ramblings about another ice age seemed preposterous, and any data surrounding this research was often taken as balderdash. Sadly, Arrhenius' theories were forgotten for a very long time, tucked away in the annals of science, locked away for many years.

And so it was, for a very long time, that climate change science diverged onto many roads that led to dead ends. In fact, until the early 1940s, many scientists thought that human influences on global warming were insignificant compared to natural forces, such as solar activity, volcanoes, and ocean circulations. It was also believed that the oceans could soak up any extra greenhouse gases that we puny humans living on such a vast planet could produce, and cancel out any pollution that we'd put into the air. In fact, in the first few decades of the 20th century, water vapor was seen as a much more influential greenhouse gas than carbon dioxide.

Research, in a way, is done at the frontier of ignorance. Scientists go in search of the unknown, persevering in the face of failure. As time went on, more and more data on the weather was being recorded, mainly from government-operated weather stations around the world. This information wasn't intended for use in ice age or global warming studies, but it did prove to be valuable in this quest nonetheless. By the early 1930s, a young meteorologist named G.S. Callendar made use of this data, and gathered temperature records from more than 200 weather stations around the world. With this data, Callendar found convincing evidence that attributed temperature increases to escalations of greenhouse gas emissions. Callendar then took his findings one step further, and pointed the finger of blame at industry—a claim that brought him ridicule from many in the scientific community and political sectors.

Callendar was an amateur in his field at the time, and many in the scientific community scoffed at his work. In particular, Callendar was blamed for missing some important factors, including cloud formations that could alter his calculations. Callendar, like most of the tenacious bunch in the scientific arena, just didn't give up. He was originally outnumbered in his battle to prove his theories, but over time, he did gain support. A handful of scientists who were skeptical of

Callendar's work, over time, were surprised to find that much of his theories were indeed correct, and that carbon dioxide could certainly build up in the atmosphere, which in turn could bring on a warming of the planet. Painstaking measurements drove home this point in 1961 when Callendar proved that the level of greenhouse gas in the atmosphere was in fact rising each year.[11]

Adding to Callendar's support were scientists who became better funded in the 1950s and 1960s from government budgets intended not to study ice ages or warming climates, but to analyze weather for battle interests of World War II and the Cold War as well. Most governments felt that anything that happened in the atmosphere and oceans could be vital to national security. So over many years, more and more scientists joined the ranks of meteorologists, and weather research gained even more momentum.

Impetus in climate change research to bolster Callendar's claims came in the form of a new tool in the mid 1950s: measurements of the radioactive isotope known as carbon-14. More commonly used back in those days as a method to date fossils and bones, meteorologists now had this tool at their disposal as well, and a chemist by the name of Hans Suess, using carbon-14 testing, found this type of carbon in the atmosphere. Suess collaborated with Roger Revelle from the Scripps Institution of Oceanography, and together, they were able to track the history of carbon-14 in the atmosphere and oceans as well. This led to even more research by well-known names in science in the late 1950s and early 1960s like Bert Bolin and Erik Eriksson of Sweden, and Charles Keeling of the U.S. The research conducted by these scientists revealed a smoking gun: a baseline for how much carbon dioxide was being pumped into the atmosphere, and how much of it nature was able to recycle back into non-threatening gases like oxygen. The findings of this research, which reinforced Callendar's claims of increasing carbon dioxide in the atmosphere, became a tipping point, winning over the confidence of more members of the scientific community at the time. Global warming had now gained a foothold in the scientific community, and news of climate change began to spread.

By the early 1970s, people in general became more Earth-conscious. Curiosity about Earth's climate turned to anxious concern, and a media-frenzy has ensued ever since. Fueling concerns about harmful invisible gases was the continual emergence of seemingly damning evidence. Of particular interest though, it was found that small, ostensibly insignificant changes could alter our environment.

WHEN TERMITES FART

Back in 1963, a mathematician and meteorologist named Edward Norton Lorenz came up with a term called "the butterfly effect." This expression turned into a common household phrase in 2004 when a movie starring Ashton Kutcher, also called *The Butterfly Effect*, was released, but had really nothing to do with Lorenz's work on chaos theory from which this phrase evolved.

Lorenz was working on the problems of complex systems, like our weather, that are difficult to predict over a given time. In the case of the weather, the period of time where we lose the ability to accurately predict anything is usually about seven days or so; TV weather forecasters are proof of that. The problem Lorenz was facing was that any model that attempts to simulate a complex system (like our weather) has to prune some information about the initial conditions. For example, when simulating the weather, you could take into account temperature from a number of areas, the wind speed, wind direction, humidity and barometric pressure. But these macro-level factors are only part of the equation, and smaller, seemingly insignificant dynamics need to be considered as well. To accurately predict the weather for more than just a few days in advance, one would need to take into account the minutest of winds from every direction that, although seemingly insignificant, can indeed have considerable impacts over a long span of time. One of these factors would be the small amounts of wind created in nature by myriad things: a breaking wave on the beach, a falling tree, and even the gentle flap of a butterfly's wings. To gather this information from every butterfly on the planet would of course be darn near impossible. But by taking into account a theory that could adjust calculations based on such small, chaotic elements, results could—in theory—become more accurate.

Addressing this problem, known in chaos theory as *sensitive dependence on initial conditions*, one of Lorenz's colleagues commented that if the theory were correct, one flap of a seagull's wings could change the course of weather forever. Later speeches and papers by Lorenz used the more poetic term *butterfly*, which sounds more angelic than *the seagull effect*. Still, the issue remained the same: a butterfly flapping its wings in Tokyo could cause tornadoes in California—theoretically speaking of course. Although this may sound far-fetched, chaos theory does prove that small changes can snowball into much larger implications. While the butterfly effect tends to refer to weather and not necessarily climate, our planet is filled with butterfly-effect-*type* conditions that can affect our warming Earth. For instance, the apparently ineffectual flatulence of a termite could in turn increase the temperature of our entire world.[12]

Termites have a rich, high-fiber diet, and they're most definitely some of the gassiest critters on the planet. You'd probably never know it, seeing as though they're so tiny and all, but when you get enough of them together, you've got one heck of a fart factory on your hands. As humorous or crude as this may sound, termite emissions contain one of the most devastating greenhouse gases: methane. What's worse, is that these tiny lumber munchers produce dangerously high volumes of it as well.

When we hear about greenhouse gases in the news, carbon dioxide (CO_2) usually takes the spotlight, and images are evoked of smoke stacks and SUVs. But there's more to greenhouse gases than just this one ubiquitous triatomic molecule called CO_2; there are quite a few vaporous substances at play here. In this section, I'll briefly describe these elements and describe their primary sources, including what happens when termites break wind.

The most abundant greenhouse gas on the planet is one that may surprise you; it's simply water vapor. Clouds do a darn good job at both reflecting and absorbing solar radiation—the heat energy that bounces back from the Earth. Among all the greenhouse gases in our atmosphere, water vapor is the most plentiful, accounting for about 95% of all greenhouse gases.[13] With water vapor being so plenteous in our atmosphere, you may wonder why this greenhouse gas doesn't get as much attention as CO_2. On this, you wouldn't be alone. Some global warming skeptics point to this statistic, and argue that the remaining 5% of supposed damaging greenhouse gases (i.e. CO_2) are insignificant. This however doesn't seem to be the case.

Water vapor has lived in harmony with Mother Nature for as long as the planet has had H_2O. In a synchronous ballet, water and air have been dancing in perfect step, and consensus in the scientific community feels that humans are doing nothing to add additional quantities of water vapor into the atmosphere. But there is some hypothesis that contrails from jetliners—those white, vaporous trails seen high in the sky as jets pass overhead—could be increasing the number of cirrus clouds in the sky, which in turn could trap in more of the Earth's heat.[14] Nevertheless, as plentiful as water vapor may be, its production by humankind is not nearly as significant as other compounds.

The second most bountiful greenhouse gas—and potentially far more devastating than plain ole H_2O—is the ever-familiar carbon dioxide, which seems to get the most airtime in stories on the evening news. Carbon dioxide is not just manmade; it's prolific in nature as well. This gas, comprising one atom of carbon, and two atoms of oxygen, is produced when any form of carbon is burned in an excess of oxygen. For example, CO_2 is released into the atmosphere during natural forest fires and volcanic eruptions.[15] Carbon dioxide is also released during the decay of dead plant and animal matter and it's also evaporated from our oceans. Even we humans release CO_2 into the air every time we exhale, as does most every other living, breathing beast on the planet. Yet CO_2 is also released during the combustion of fossil fuels like the gasoline that powers our cars, and coal that's fueling many of our nation's electric plants.

Luckily, carbon dioxide is removed from the atmosphere by what are known as carbon "sinks." These sinks include such things as absorption by the seawater in our oceans, and photosynthesis of ocean-dwelling plankton and land-dwelling biomass such as forests and grasslands that, unlike us humans, breathe in carbon dioxide and exhale life-giving oxygen. Interestingly enough, some of the best carbon sinks are tiny little, single-celled marine plants known as coccolithophores. These small organisms that are highly prolific around our world's vast oceans, love carbon dioxide. These tiny plankton-type creatures breathe in carbon dioxide, dissolve it, and create a calcium byproduct to form their crusty shells. Incredibly enormous populations of these tiny organisms are constantly at work around Earth's oceans, merrily digesting CO_2 to make their brittle calcium coats of armor. After a very short life, the coccolithophores die and fall to the bottom of the ocean. On the ocean floor, they form sediments of limestone and chalk, and eventually, tectonic activity raises them above sea level to form giant cliffs, where common everyday schoolroom chalk is often excavated. It's hard to believe that the breathtaking views of the White Cliffs of Dover on the coast of England were created by one of the world's best carbon sinks, tiny little coccolithophores—but it's true.[16] Mother Nature has a great way of recycling waste—by turning trash into treasures, refuse to riches, and carbon dioxide into crusty little coccolithophores.

Unfortunately, carbon dioxide is released into the atmosphere by many unnatural things, including gasoline powered vehicles, iron and steel production, the manufacture of cement, ammonia production, and natural gas flaring to name a few. This addition to Earth's naturally occurring sources of carbon dioxide makes it difficult for Mother Nature to keep up with her recycling program. It's estimated that nearly 24 billion tons of CO_2 are released into the atmosphere each year from the combustion of fossil fuels alone.[17] Now although this may sound like a colossal quantity of carbon dioxide, it's really just a fraction of the total 600 billion tons released into the atmosphere each year, much of which is from natural sources. Moreover, total human emissions of CO_2 annually only account for about 5% of the total amount of CO_2 emissions, or roughly 30 billion tons in all.[18]

Nevertheless, our seemingly small 5% addition to Mother Nature's CO_2 scales has tipped her balance in the wrong direction. Carbon dioxide sinks have been unable to keep up with this excessive outpouring of CO_2, leaving us with some fairly hefty atmospheric concentrations of CO_2 that just won't go away and continue to accumulate in the sky above us. When measuring the concentration of any type of gas in the atmosphere, scientists use a unit known as "parts per million," or just ppm for short. This unit of measure signifies one part of something compared to a million parts of something else. In the case of greenhouse gas, 1 ppm means that 1 unit of the gas is present for every 1 million units of air. One ppm is the same as 1 minute in 2 years (which is a little over 1 million minutes) or 1 cent in $10,000. When talking about atmospheric concentrations of CO_2, recent measurements show that we have around 372 ppm of CO_2 now, compared with only 280 ppm before the onset of the Industrial Revolution[19], for a total increase of nearly 33%, and rising.

So although water vapor is big man on the greenhouse gas campus, CO_2 is becoming more abundant, and is thus causing a warming type of butterfly effect. What's especially worrisome though is that it's believed that the human-induced 30 billion tons of CO_2 released today into the atmosphere will rise in coming years as populations grow and world economies flourish. It's estimated that in the year 2015, human emissions of CO_2 will increase to over 36 billion tons annually, and by 2025, we'll be pumping out nearly 42 billion tons of CO_2 a year.[20]

The number two henchman in the greenhouse gas gang is that offensive asphyxiant emanating from those pesky wood-eating critters that we call termites. This volatile vapor is methane, also known as CH_4. Although methane is released in much smaller quantities than carbon dioxide, it's over 20 times more effective at trapping heat in the atmosphere than CO_2, which is why it is of great concern despite the fact that it's released in smaller quantities than its bigger cousin: carbon dioxide.[21]

Methane comes from a variety of natural and human-related sources, yet human-related sources comprise the majority of methane production, topping the scale at an astonishing 60% of all methane emissions worldwide. Natural sources account for only 40%.

Of the 40% of natural methane sources, wetlands top the list. Natural wetlands, like Florida's Everglades, are responsible for approximately 76% of global methane emissions from *natural* sources, accounting for about 145 million tons of methane per year. Wetlands provide a habitat conducive to methane-producing bacteria that generate methane during the decomposition of organic material. Things slowly rot in swamps, marshes, and bogs, and a methane byproduct is released as a result.

The second leading source of natural methane emissions comes from those notorious timber-munching termites. As implausible as this may seem, farting termites account for 11% of global methane emissions from all natural sources on our planet. These tiny lumberjacks break enough wind to emit nearly 20 million tons of methane gas into the atmosphere each year. Methane is produced in termites as part of their normal digestive process. Like we humans, when ingesting copious quantities of fiber, our digestive system produces vaporous byproducts, which are—to the disapproval of immediate company—released into our surrounding area, and eventually skyward. With termites, their emissions can be so powerful that some pest control companies actually use dogs, trained in the art of gas whiffing to sniff out these tiny little lumber destroyers.[22] As a side note though, before you try to pass blame on termites when being accused by someone who's become olfactory-offended by your discharge of unpleasant effluvium, bear in mind that methane is not only colorless, but odorless as well; the offensive aromas associated with digestive-originating miasmas are a result of other compounds made during the digestion process—not methane.

The remaining 13% of natural methane emissions come from our oceans and what are known as hydrates, usually found in ocean sediments. Although the source of the non-hydrate ocean-originating methane emissions is not yet entirely understood, it is believed that as with termites, small zooplankton and some species of fish emit methane as a byproduct of their digestive process.

Obviously, there's a lot of methane being produced by Mother Nature, but as mentioned, this ole gal only creates about 40% of this precarious greenhouse gas. The other 60% are from manmade sources including landfills, natural gas systems, coal mining, wastewater treatment, rice cultivation, petrochemical production, and manure management to name a few.

Landfills are the worst offenders though, accounting for 34% of all human-related methane sources. Landfills slowly decompose our trash, and in the process, methane is emitted into the air. Natural gas systems rank up there as well and run number two on the manmade methane hit list. Methane is actually the primary component of natural gas itself, but a good amount of this gas is lost during the production, processing, storage, transmission, and distribution of natural gas. Because natural gas is often found in conjunction with oil, the production, refinement, transportation, and storage of crude oil is also a source of methane emissions.

The third highest source of humanly responsible methane emission comes from none other than creatures breaking wind. Unlike termites or zooplankton though, these animals are domesticated livestock, and are thus considered to fall under the category of human-related methane sources. The biggest troublemakers are our milk-manufacturing friends: cows. Falling under the class of animals known as *ruminants*, cattle have a complicated digestive system, which includes four stomachs. In the rumen, or large fore-stomach of these animals, microbial fermentation converts grass and feed into products that can be digested and utilized by these behemoth bovines. Grass isn't the most nutritious substance, but it is plentiful. Still, cows have to go through a slow, meticulous process to break down the grasses and feeds they consume, and turn this into sustenance for their massive physiques. The microbial fermentation process that occurs in their rumens, often referred to as *enteric fermentation*, produces methane as a byproduct, which is expelled by either burping, or when they pass gas.

Cows produce about 80 million tons of methane annually around the world, accounting for about 28% of global methane emissions from human-related activities.[23] An adult cow may be a very small source by itself, emitting only 80-110 kilograms of methane, but with about 100 million cattle in the U.S. and 1.2 billion large ruminants in the world today, cow-like creatures are one of the largest methane sources around. In the U.S. alone, cattle emit about 5.5 million tons of methane per year into the atmosphere, accounting for 19% of human-related methane emissions in the United States.

Manure is also a big contributor to this methane problem. Since decomposing matter has the nasty tendency to emit methane, cow dung, and the massive quantities of it, can be problematic to say the least. Although manure deposited on fields and pastures—or otherwise handled in a dry form—produces insignificant amounts of methane, liquid manure management systems, such as lagoons and holding tanks commonly used at large swine and dairy operations, can cause significant methane production.

But cows aren't the only sizeable creatures populating our planet that are creating methane-emitting waste material; we humans do so as well. Wastewater from municipal and industrial sewage is treated to remove soluble organic matter, suspended solids, pathogenic organisms, and chemical contaminants. These treatment processes often produce methane emissions. In addition, the sludge produced from some treatment processes may be further biodegraded, resulting in even more methane.

Methane, like carbon dioxide, also has sinks that Mother Nature uses to eliminate excessive quantities of this gas. The atmosphere, and more precisely the troposphere, is the largest sink for methane. Methane in the troposphere reacts with a highly effective and naturally occurring compound in the Earth's atmosphere known as hydroxyls. This methane sink accounts for nearly 90% of all methane recycling, which turns this bothersome greenhouse gas into mostly water vapor, and another compound known as CH_3. Earth's dirt helps to eliminate most of the rest.

Even though methane is not the most abundant of greenhouse gases, when comparing it to carbon dioxide, it does a far better job at holding in heat, and with methane levels now 2.5 times greater than they were before the Industrial Revolution, there is grave concern of how to deal with it.[24]

There are other greenhouse gases at play as well including nitrous oxide, emitted during agricultural and industrial activities and during combustion of solid waste and fossil fuels. Hydrofluorocarbons (HFCs), perfluorocarbons (PFCs), and sulfur hexafluoride (SF6), are also added to the atmospheric cocktail of heat trapping gases in our sky, which are generated in a variety of industrial processes, and do not occur naturally on their own.

There's no doubt that levels of *all* greenhouse gases have increased over time. Concentrations of heat-trapping gases are on the rise. But what—if anything— have they actually done to harm our environment?

THE FIRST SIGNS

Easter Island was once richly forested with a diversity of woodlands and wildlife, but it is no more. This now, nearly barren island sitting about 2,000 miles from Chile, and another 2,000 miles from Tahiti, is best known now for its gargantuan stone head statues called Moai, but no longer its trees. Over a slow progression that lasted a few centuries, the people of Easter Island decimated their homeland by wiping out every single tree on the island, which in turn drove other plants and animals to extinction, forcing their complex society to spiral into chaos and cannibalism.

Pulitzer Prize winner Jared Diamond, Professor of Geography and Physiology at UCLA coined a new term in his book *Collapse* called "landscape amnesia." Also known as "creeping normalcy," Diamond uses this expression to explain the mentality of the Easter Islanders. A question was posed to Dr. Diamond by one of his students, asking what was going through the mind of the Easter Islander who cut down the island's last standing tree. As Diamond explains, since it took hundreds of years for Easter Island to befall its deforested state, each new generation saw things only slightly different from the last. The changes observed by a son or daughter were barely any different from what their parents had witnessed during their lifetime. As time went on though, and many generations had passed, the overall change had become significant; however, each generation, observing only minor changes on the landscape, practically forgot how things used to be.

Easter Island's demise was a prime example of landscape amnesia: forgetting how different the surrounding landscape looked many years ago, because the change from year to year has been so gradual.

Our Earth has gone through changes since its inception. Our temperature has not always been the way we know it now; things were quite balmy when dinosaurs were trampling about, and yet the climate was quite nippy when wooly mammoths were nearing extinction during the last ice age. Even now we're going through changes, but since our planet has changed so radically over its 4.5 billion year history[25], we have to wonder if the recent changes we're seeing now are from the greenhouse gas emissions created by humankind, or is this something that is naturally occurring? In any case, there is a bit of landscape amnesia going on. Many things have slowly changed over time, leaving a view of a past seemingly similar to the present, yet slightly different than now. In this section, I'd like to point out a handful of changes,[26] which seem to be linked to global warming—whether caused from the hand of man or otherwise—that exhibit an eerie kind of creeping normalcy.

Earth's average surface temperature has increased by about 1°F over the 20th century[27], with polar regions seeing increased temperatures of more than 2°F. Although this minor tick of the mercury would hardly be noticed by even the most sensitive among us, Mother Nature—being the temperamental ole gal that she is—is feeling the heat. During the time of this increase in temperature, the average global sea level has increased at a rate of 1-2 millimeters each year, and the duration that ice remains frozen on our rivers and lakes during the winter has declined by about two weeks. This information, provided by the Intergovernmental Panel on Climate Change (IPCC) is backed up by photo-graphs taken in Alaska around 1950 to pictures taken around 2000. These pic-tures, comparing that 50-year span, show an increase in shrub abundance in tundra areas and an increase in the extent and density of spruce forest along the tree line. Also, in Alaska's interior, permafrost thawing is now causing the ground to subside by as much as 33 feet in some areas.

Glaciers are also melting. A study conducted in 1999 shows that between the mid-1950s and mid-1990s the glaciers around Alaska thinned by an average of about 1.6 feet each year for 40 years, and then repeat measurements showed that from the mid-1990s to 2001, the rate of thinning had increased to nearly 6 feet per year. The Canadian Rockies' Athabasca Glacier has retreated one third of a mile in the last 60 years and has thinned dramatically since the 1950s. In British Columbia, the Wedgemont Glacier has retreated hundreds of meters since 1979 as well, and the Bering Sea and Arctic oceans are seeing a dramatic reduction in sea ice.

Weather seems to be affected and heat waves have become more frequent. Texas endured a deadly heat wave in the summer of 1998, claiming more than 100 lives in the region. Dallas temperatures were over 100°F for 15 straight days. That same year, Melbourne, Florida suffered 24 days above 95°F while nighttime temperatures in Tampa remained above 80°F for 12 straight days. During the fall of 1998, an unprecedented autumn heat wave from mid-November to early December broke or tied more than 700 daily-high temperature records from the Rockies to the East Coast.

Then in 1999, New York City had its warmest and driest July on record, with temperatures climbing above 95°F for 11 consecutive days—the most ever in a single month. The month before however, more than 250 people died as a result of a heat wave that gripped much of the eastern two-thirds of the country. Heat indices of over 100°F were common across the southern and central plains, reaching a record 119°F in Chicago. And in recent days, the heat wave of 2003 caused thousands of deaths across Europe, with some estimates pointing to as many as 35,000 to 50,000 deaths from the abnormally hot summer temperatures.[28]

Species around the globe seem to be disturbed as well. In Canada's Western Hudson Bay, adult polar bears have been decreasing in weight, and there's been a decline in birthrate since the early 1980s, which has been attributed to the earlier spring breakup of sea ice. Alaska's sea bird population has seen recent declines with the black guillemot population declining due to melting sea ice that has increased the distance the birds must fly to forage for food—reducing the number of resting sites available at the same time. Caribou have been dieing off as well. Peary caribou in the Arctic have declined from 24,000 in 1961 to perhaps as few as 1,100 in 1997, attributed primarily to freezing rain covering the animals' food supply.

In California, shoreline sea life has been shifting northwards. Changes in invertebrate species such as limpets, snails, and sea stars in the 60-year period between 1933 and 1993 indicate that species' ranges are shifting northwards, which is thought to be in response to warmer ocean and air temperatures. And down in the Monteverde Cloud Forest of Costa Rica, frogs and toads have been disappearing due to a reduction in dry-season mists from a now warmer Pacific ocean. Recent studies also point to declining frog populations in Central America from more prolific fungi.

Plant life seems to love all this warming though. Cherry trees around Washington, D.C. are now blossoming earlier than ever. The Olympic Mountains in Washington State are seeing a proliferation of alpine meadows, where the sub-alpine forest has invaded higher-elevation alpine meadows, partly in response to warmer temperatures.

Disease and insects however are becoming worrisome. In North America, ecologists have identified the first genetic adaptation to global warming in the North American mosquito *Wyeomyia Smithii*. Modern mosquitoes now wait nine days more than their ancestors did 30 years ago before they begin their winter dormancy, with warmer autumns being the most likely cause. Higher temperatures are enhancing mosquito survival rates, population growth and biting rates, which can in turn increase the risk of disease transmission. And in Mexico, Dengue fever has spread above its former elevation limit of 3,300 feet and has appeared at altitudes as high as 5,600 feet.

These are just a few of the first signs that our planet is undergoing change. Much more is afoot, which I'll discuss in more detail throughout the next few chapters. Yet it is important to note that these signs, which often fall victim to a kind of landscape amnesia, have occurred just recently.

How Revolting

Back in the mid 1990s, my sister's daughter Abby was on the verge of becoming a teenager. Prior years were easy on her uncle's Christmas present budget, and she'd be overjoyed upon receiving inexpensive toys, stuffed animals and such that I'd bring as gifts. As Abby grew older, her wants became a bit more expensive. During one visit to my sister's home, Abby was begging for a fax machine, which her parents insisted would not just fall out of the sky nor grow on any nearby tree.

Abby persisted and whined a bit, pleading with her parents to buy her this magically magnificent wonder of communication. It was the *in thing* with her peers, and all the rave amongst her classmates. If you had a fax machine, you were cool. When trying to rationalize why she needed a fax machine, Abby's reply was simply, "...everybody else has one," which was true—to a certain extent. I had one in my office at home, and those who received my facsimile transmissions had one as well. Nevertheless, her argument was not compelling enough, and much to her disappointment, Abby never got her fax machine that year.

Fast-forwarding into the new millennium, Abby was now looking into colleges she'd be attending, and she didn't seem to be too mentally scarred from living without a fax machine during her high school years. Jokingly, during one visit around 2001, I asked her if she'd like my old fax machine. I didn't need it anymore; email was the *de facto* standard form of communication, and replaced the need for paper. I'll never forget the look I got from my niece. Her expression radiated a "get with the times old man" message, as though I was wearing tie-died tshirts with bellbottom trousers, wondering where I placed my favorite *Saturday Night Fever* 8-track tape.

Abby didn't need some bulky fax machine to tie up the phone line at her house. All she had to do was whip out her cell phone, and either call or text-message her friends. And if she wanted, she could just log into her computer and send off an email. It's funny looking back on the "old days" before cell phones were ubiquitous and the Internet an everyday household utility, but what's even more interesting is that it wasn't all that long ago. Only a few years had passed between Abby's desire for a fax machine to the time she got her first cell phone. During that short span of only a few years, communication and technology had changed dramatically.

It's mind-boggling to think that only in the last 130 years have we gone from telegraphs to telephones, and now cell phones and email. Modern humans went without such luxuries of communiqué for tens of thousands of years, and then all of a sudden, technology advanced at blazing speeds. The spark that drove all of this was the Industrial Revolution that began in Britain in the early 18th century, and started in the United States shortly thereafter. This was the tipping point that set humankind on a speedy treadmill running ever faster toward new discoveries, giving us the power to collaborate and invent like never before. Empowerment first came in the form of mechanization that turned manual labor from homes and small workshops into manufacturing in large factories. It was a time of exploding ideas with new emerging resources to not only make things faster, but to discover new ideas as well. There became a greater availability and variety of material goods, which accelerated over time. This has helped to fuel advances in technology that today, provide us with many luxuries in transportation and communication that we oftentimes take for granted. All of these advancements however have come at a cost.

The Industrial Revolution, in its incipiency, brought on new ways to make iron, using the more efficient and inexpensive process of burning coal. Iron could now be made faster, which is partially responsible for the rapid developments of railroads, steam engines, and engine building in general. Factories took form, and increasingly more products were being turned out at amazing speeds. Although transportation and manufacturing processes were advancing, and iron was being hammered out faster than it ever had before, the new industrialized processes were now pumping wasteful byproducts into rivers, streams and lakes, and unfortunately, the Earth's atmosphere. This trend hasn't stopped, and has continued to this day.

All measurements require a baseline. You need a reference point to measure point B to something else (typically point A). When measuring how tall a skyscraper is, our reference is the ground. When measuring how high a mountain is, we use sea level as our point of relativity. And when analyzing global warming, the Industrial Revolution serves as one of our datum lines. When talking about carbon dioxide levels in *When Termites Fart*, I briefly mentioned the difference between CO_2 levels now when compared to the Industrial Revolution, and throughout this book, this point of reference will be mentioned from time to time as well. The Industrial Revolution was a turning point in human development. Yet since this also marks the beginning of increasing use of fossil fuels and subsequent greenhouse gas emissions, the Industrial Revolution is also one of many important epochs in the science of global climate change.

Humankind has only seen a smidgen of the Earth's history. We've barely glimpsed a blink in time. This old planet we're sitting on has been around far longer than we have. In fact, if we were to look at the life of our planet up to now as the length of one mile, the existence of modern humans wouldn't even take us back a half of an inch. And when thinking of how long it's been since the Industrial Revolution, you'd have to use a magnifying glass and a much smaller ruler to measure our blip of only about 0.003 inches along Earth's one-mile time-line.[29]

During our brief stay on this bluish green planet, we've excelled significantly as a species, yet we don't know entirely what life was like way back in the first few yards of that mile-long time scale showing the ticks of time and the advancement of the ages. There are certain markers along the way though that have been discovered, and just how that relates to global warming is surprising.

Throughout this book I'll be discussing what we do know about Earth's climate throughout history and what we can tell from atmospheric readings and other data that tell a story of Earth's *climatic* past. Nevertheless, human history recorded over the past few thousand years or so does provide some reference to how landscapes looked, and how people lived during these times. We also know that even during our blip on Earth's radar of time that these things have been changing recently, and as a result, it's affecting people today, and in some unusual, unsuspecting ways.

2

Selling Sunscreen to Eskimos

○ ○

The Sun, with all those planets revolving around it and dependent upon it, can still ripen a bunch of grapes as if it had nothing else in the universe to do.
—*Galileo*

When global warming stories make the news, we'll often hear of impacts on glaciers, ice caps, wildlife, and other such things in nature. Declining polar bear populations, interrupted bird migrations, intense hurricanes and heat waves occupy most of the headline real estate on front-page newspapers. While these issues are real and newsworthy, they frequently lack the human touch, making it sometimes hard to realize just how serious the problems are—and can become—in a heated world. Rarely do we hear how our changing climate is altering lifestyles of people around our planet. In many respects, some of this is due to landscape amnesia, where small changes have occurred over such long periods of time and the apparently tiny transformations observed in recent years hardly seem out of the ordinary. Moreover, as a species, we humans have done a darn good job at adapting to our surroundings no matter what nature may throw at us. Early Europeans learned to don thick fur coats in the summer tens of thousands of years ago during the last ice age. Our ancient ancestors also learned to cultivate food and raise livestock to ensure survival in an often-unpredictable hunter-gatherer world. And even in the last century, our technological advancements have allowed us to bring water to desert locations like Nevada and Arizona to build great cities where once there was nothing but dry, arid desert and wasteland. Human survival, mixed with slow changes of creeping normalcy can lead to a feeling that nothing has ever changed—but it has, and still is.

We often think of our planet as a static entity, steadfast and unchanging—but it's not. Our world has and will continue to go through many transformations through its life in the cosmos. Some recent adjustments though have been sudden, thus ringing alarms that something is awry. Still, quite often we tend not to notice how the world itself may be changing, and in turn altering human lifestyles as a result.

For example, living in Southern California, I take for granted the bottles of 45 SPF sunscreen kept in my car and medicine cabinet at home. I use this lotion on a regular basis to keep my delicate, ivory colored skin from harm. I can go to any drugstore or supermarket in my neighborhood and pick out a variety of this funny smelling skin goop to defend my dermis. It's a ubiquitous over the counter ointment in sunny Southern California—and much of the lower 48—that costs but a few bucks per bottle. Yet I should count my blessings that I live around the latitude of 34° north, and not in the Arctic.

Sunscreen is an import to the people occupying the far northern reaches of our planet. In the past, it was never needed. Only recently have cases of skin cancer affected Inuits of the north, making sunscreen now a valuable commodity to a people who for generations lived much of their life outdoors, not fearing protection from the Sun. But in a world that's warming, lifestyles are changing, in a life-or-death way. Inuits of Canada are now finding sunscreen to be as valuable as mukluks and parkas, and has become a part of their regular, everyday lives to protect their health. Unfortunately, sunscreen is also expensive in the Arctic regions, and as its popularity grows, supply and demand continues to force sunscreen prices ever higher. With the Sun now threatening Inuits with melanomas, new government programs have emerged for those unable to purchase the expensive skin protectant.[30]

Eskimo sunscreen is a sign of our times; it's just one of many indicators that our world is now changing. Climate change is most definitely upon us, this we know. We know the Earth has been warming lately, and even the most ardent global warming skeptics agree on this point. The crux of the global warming argument stems from whether these changes in temperature are due to human-induced causes or just natural fluctuations, thus the fact remains that we do know that global temperatures have risen, and there are indeed transformations taking place around the world—no matter what the root cause may be.

In this chapter, I'll discuss how global warming—whether human-induced or not—is altering the way some people now live, and how their surrounding environments are changing as well. In a world where it's become easier to sell sunscreen to Eskimos, many other perplexing events are taking place.

SURFING THE TUNDRA

Zoya Telpina is just one of about 350 Chukchi people living in a tiny port village known as Yanrakynnot, located on the rugged shoreline of Russia's Chukotka Peninsula near the Bering Strait. Zoya has lived in this icy realm of the Eskimo all of her life, and serves as the local schoolteacher to children of the village, where reindeer herding and whale hunting is the primary source for survival.[31] Hers is a simple existence, one that's been passed down from generation to generation, from elders in tune with the nature that brings forth their bounty. The Chukchi people have learned through the ages how to read the sky, thus knowing with an amazing degree of accuracy when to embark on a hunt, or when to avoid blizzards looming on the Arctic horizon. But this is now changing, and in ways the Chukchi people find hard to explain.

In the Arctic, the ground is known to remain frozen throughout the year, and thus aptly named *permafrost*. But the once frozen ground is now coming to life. The ice that blankets winter seas around the Bering Strait has always been so thick that it could easily support Chukchi men dragging their sleds loaded with whale, walrus, and seal carcasses returning from a hunt. But that's now changing too. Last winter, when Zoya Telpina looked out from her kitchen window toward the Bering Sea, she saw something she had never seen before in all her years: the ominous blue of the open ocean, surf at the edge of the tundra. Water now ebbed and flowed where there had always been solid ice. Zoya's Arctic home, much to her chagrin, had become a beachside residence, living on the edge of the pounding surf that, during heavy winter storms, could easily erode away the coastline surrounding her village.

Zoya is lucky though to be situated high enough from sea level—for now—to keep her home, yet she may soon meet the fate of other nearby Inuits. Sadly, many of Zoya's neighbors have already been forced to flee their homes. In late 2005, the 600-person Inupiat Eskimo community of Shishmaref, located on the Chukchi Sea just north of the Bering Strait had no choice but to leave their ancient village as pounding surf—no longer held back from the protection of winter sea ice and permafrost—eroded their homeland.[32]

The impacts on Arctic people have become quite complex. In a way, their entire ecosystem is deteriorating before their eyes. Between 1978 and 2005, Arctic sea ice has been reduced by 24% in size, and its thickness only 50% of what it once was.[33] As this sea ice melts and relinquishes to the sea from whence it came, the remaining thinner ice becomes a hazard. This precariously thin sea ice is now a dangerous passage for hauling mammal carcass-laden sleds, and thus inhibits hunting expeditions to areas once rich with game. This ice thinning is hindering gillnet fishing as well, since it has become a game of chance to venture out onto the edges of the ice to approach the sea and cast in nets.

Once Eskimo and Inuit people do find safe passage across the melting ice, they discover that their hunting grounds are now becoming scarce with less food than previously available. Polar bears are becoming less prominent and with numbers in obvious decline, U.S. Fish and Wildlife officials in February 2006 began to review whether to put polar bears on the endangered species list.[34] An estimated decline of 15% of the polar bear population around Hudson Bay, and sightings of thinner, more aggressive bears prompted much of this action. As disheartening as this may be to animal lovers, it's even more disturbing to the people of the north who rely on polar bear hunting for not only food and warmth, but for commerce as well. Changing climate has affected these bears in similar ways as the people of the frozen north that hunt them.

Currently, it's estimated that there are about 22,000 polar bears living around the Arctic Circle. These great white beasts spend much of their lives hanging out on the vast floating sea ice where they sleep, mate and hunt for prey, such as seals. But in the past few decades, warming temperatures have caused the sea ice to break up earlier than normal, well before spring arrives. And worse, the ocean is freezing over later in the fall, thus extending the warmer summer season. In 2004, scientists witnessed four drowned bears floating in open water, apparently exhausted while trying to swim a futile 180 some miles between ice and land in high winds—a distance far greater than these bears are accustomed to.[35] The thinning and subsequent cracking of the pack ice is also forcing polar bears to come onto dry land to search for their food. This isn't a good thing. Since seals have always ranked high on the polar bears' menu, finding sustenance on land is usually an unproductive task. Unfortunately, as bears forage for food on land, and after failing to successfully fill their ravenous appetites, they get quite cranky, and become a danger to anyone within smelling distance.

Polar bears' main prey, seals, have also been taking a hit from the heat. Declining sea ice has forced grey seal populations to birth their pups on land and not on ice floes as they usually would. This leaves the seals vulnerable to conditions that they're unacquainted with. In February 2006, numerous seals, unable to find stable ice floes, took to Pictou Island off of Nova Scotia as their port in the storm, which they sadly found out, was not as safe a harbor as the sea ice they would usually perch upon. In February 2006, with hundreds of seals sprawled out on the shores of Pictou, tending to their newborn pups, a massive storm rose up and ravaged the coast of the island, attacking the defenseless mammals not habituated to life on land. As waves pounded the shores of Pictou Island, the sea rose up to claim its own. Storm surge and rising tides amongst the treacherous surf washed hundreds of newborn seal pups out to sea, and to their deaths.

Witnesses to this tragedy, helpless to assist the young seal pups, watched in horror as mother seals put forth stoic yet hopeless attempts to push their young back to shore with their dog-like snouts. Adult seals were seen diving in vain to lift the drowning pups up from the briny deep, but merciless waves continued to pummel the coastline, sweeping the poor pups back out to sea, and to a watery grave.

Other mammals in the Arctic region have taken a blow from the warming climate as well. The suspicions of many Inuits that whale populations are declining, and that these leviathans of the deep are becoming thin, is bolstered by recent reports recording waning numbers of bowhead, gray and right whales. In the last few years, Eskimo whalers talked of harpooning "stinky" whales that appeared to be rotting alive, with their meat so smelly that the Eskimo's pack dogs snubbed their snouts at it. And residents as far south as the Puget Sound, San Francisco, and even Baja have seen emaciated gray whale carcasses drift toward shore, already dead.[36] It was thought that perhaps pollution or a mammalian disease could be the culprit, but studies suggest this is not the case. Many experts now theorize that these whales are finding their food sources hindered by recent increases in sea surface temperatures[37], and in many cases are literally starving to death. Rights, grays and other such whales, sustain their massive physiques by gorging on immense quantities of tiny shrimp-like creatures known as krill. Although usually quite abundant throughout the world's oceans, it's believed that krill are now either declining in numbers or relocating from the recent warming of ocean waters, making it tough for hungry whales to find a bountiful banquet of theses tiny crustaceous amphipods for their sustenance.

The top of the food chain, the people of the Arctic, are now suffering from a dwindling supply of their natural food sources such as whale blubber and seal meat. Living in a climate conducive to hunting and gathering but not farming, when the mammals disappear, so does a valuable protein source for the indigenous people of the north. It is argued by some global warming skeptics that increased temperatures in the Arctic may actually help the food chain by increasing the plankton in the ocean water, which would give fish more food, which in turn would feed more seals, thus giving polar bears bigger, juicer seals to gnaw on.[38] So far, this doesn't seem to be the case.

Although carbon dioxide is a wonderful fertilizer, and plants thrive on the stuff—as do plankton—the increases in CO_2 so far do not seem to be doing as much good up in the great white north as the melting ice is doing harm. If seals are hindered in their migration or breeding cycles, and polar bears can't get to them, then increases in plankton do little good. And if krill, which feed on phytoplankton and zooplankton, are decreasing or relocating, then any useful fertilization effects from CO_2 are yet to be realized. Instead of CO_2 helping the situation as some skeptics suggest, it seems to be adding insult to injury. Inuits are now finding their mammalian prey diminishing in number, dwindling in size, and increasingly more difficult to access, since the food chain's most fundamental link has been disturbed.

For now, the indigenous people occupying our planet's uppermost latitudes will need to adapt to the changes before them. With more of their natural food supplies diminishing, more Inuit and Eskimo people are forced to rely on more expensive, store-bought food, which in turn may damage their diets and overall health as they consume cheaper, high carbohydrate foods richer in refined sugars, high fructose corn syrup, and other high calorie, low protein fare.

Yet the people of the Arctic, like many generations of human populations before them, are doing their best to acclimate to their changing world. In some cases, sled dogs are being bred with shorter haired hounds to produce a pooch that's able to run in the warmer temperatures and not exhaust as quickly as long-haired huskies or malamutes. And with the permafrost now thawing, more ground is becoming available for growing some crops, thus allowing more Inuits than ever to grow gardens of potatoes, cabbage, turnips and peas.[39]

Besides dwindling food supplies and loss of hunting grounds and safe passage, people of the north have other concerns from melting sea ice. In the last three decades, warmer Arctic temperatures have caused a 25% decline in sea ice near the Northwest Passage that links the Atlantic to the Pacific Ocean through the Arctic archipelago of Canada.[40] In the past, this region was always so frozen-over that only specialized ice cutting ships could make the perilous passage. But with less ice present now that would normally hinder the crossing of commercial vessels across the Northwest Passage, there is worry that more ships may use this route, thus increasing the risk of maritime disasters like oil spills. With the Northwest Passage becoming a safer body of water in which to navigate, an increase in commercial fishing could be brought to the region as well, which could further decimate local fish populations. Canadian national security would also be a concern and increases in naval protection may be required over time.

Mosquitoes are another fear. The Arctic has an extremely heavy population of these tiny bloodsucking insects where millions can explosively emerge from damp, warming tundra during the summer in a single day. Myriad mosquitoes are nothing new to those who live near the Arctic Circle, where during warm weather, clouds of mosquitoes can turn the sky gray. Yet recent increasing temperatures seem to be bringing on higher numbers of mosquitoes during the summer—a season now lasting longer than before. This is not only a nuisance, but also increases the risk of blood transmittable diseases such as yellow fever, dengue, encephalitis, Rift Valley fever and West Nile virus. A study in 2001 also confirmed that mosquitoes are now genetically adapting to warming climates, with one species in particular, the pitcher-plant mosquito *Wyeomyia smithii*, waiting an extra nine days to begin its winter dormancy.[41] Many Inuits claim that mosquitoes have become more prolific, thus raising their concerns of health and safety.

Of all the alarming interests of a warming world though, the main concern of most Arctic people is their way of life and food supply. After living in a region for a thousand years or more, many indigenous people of the Arctic are finding adaptation difficult. A changing climate, whether from the deeds of humankind or not, is transforming the Arctic landscape and the lifestyles of the people that occupy these high latitude regions. Inuits are well aware of the problems surrounding them, and many are pointing the finger of guilt at human-induced global warming. The Inuit Tapiriit Kanatami people of Northern Canada have taken a stance on global warming being human-induced, and regularly publish articles related to climate change in their Environment Bulletin reports. Additionally, an organization known as the Aboriginal and Northern Community Action Program (ANCAP) has taken a similar stance, and has set forth objectives to reduce greenhouse gas emissions.[42]

In a fight to protect their right to ice, the Inuit Circumpolar Conference (ICC), a federation of native nations representing about 150,000 people in Canada, Greenland, Russia, and the U.S., filed a petition with the Inter-American Commission on Human Rights (IACHR), blaming the U.S.—a major contributor of greenhouse gases—for the Arctic's rising temperatures and injured ecosystems. The ICC tried a similar move back in 2003 at the ninth annual meeting of the nations and nongovernmental organizations negotiating the emission-reducing plan known as the Kyoto Protocol.[43] The ICC's petition in 2003 went over like a lead balloon, but the ICC refused to give up, and submitted their petition in December of 2005.[44] This 163-page petition, supported by testimony from 63 Inuits in Canada and Alaska, urges the IACHR to recommend that the United States adopt mandatory limits to its emissions of greenhouse gases and cooperate with the community of nations to prevent human-induced interference with the climate system. The petition also requests that the IACHR declare that the United States of America has an obligation to work with Inuit to develop a plan to help Inuit adapt to unavoidable impacts of climate change, and to take into account the impact of its emissions on the Arctic and Inuit people before approving all major government actions.

The ICC petition is a major step for the Inuit people, provided it can be proven that the global warming impacts being felt in the north are indeed caused by human deeds, and not natural causes. After all, the Earth is obviously warming up, but who's to blame? While the devil's in the details that I'll be covering later on in chapters 4 and 5, it would seem logical to find heavy greenhouse gas emitters like the U.S. responsible for Arctic transformations, yet an injustice would be served if science proves otherwise. In either case, it could still take another two years before the IACHR can make a determination on the matter.

In the meantime, in valiant efforts to defeat the onslaught of increasing greenhouse gas emissions, stoic members of the Inuit society are doing all they can on a personal level. One man, Simon Hohlmeister who's lived near the Newfoundland municipality of Nain, Nunatsiavut for most of his life, has mothballed his gas-burning all terrain vehicle (ATV), and reverted back to a dog sled team for his primary transportation, fearing the fumes of his ATV will further harm the environment and decimate his people's way of life. Sadly, Simon's efforts and ATV emissions pale in comparison to the plenteous amount of greenhouse gases that are emitted by his industrialized neighbors to the south. Still, with little else he can do, Simon and many like him in the north feel they are doing what they can to save what they have left.

HEY, WHERE'D THAT ISLAND GO?

The island nation of Kiribati, comprising 30 some South Pacific atolls and islands, is a tropical paradise situated just to the southeast of the Marshall Islands on the balmy latitude of the Equator. Uncrowded, white sand beaches greet clear blue water with a lush green backdrop of palm trees—truly the quintessential vision of an island utopia. Recent changes in the world's climate however are now transforming this Shangri-la of the South Seas.

Navigating around Kiribati's waterways, islands and atolls changed around 1999 when tragedy struck two small islands known as Tebua Tarawa and Abanuea. Although uninhabited, and only used by local anglers as quick stopovers on their fishing routes, these two small islands were part of Kiribati's republic, serving as a landmark in the similar way that the Channel Islands off of California do; they just sit there as part of nature, accenting the beauty of nearby inhabited land miles from their pristine shores. Unlike the Channel Islands though, Tebua Tarawa and Abanuea were never very tall, with elevations barely above sea level—a target for disaster in a world where ocean waters are warming and sea levels are rising. Nowadays, the tiny islets of Tebua Tarawa and Abanuea are no more. They have completely disappeared. The sea rose high enough in the 1990s to submerge these small islands, which seems ironic, since the Kiribati word for Abanuea means "the beach which is long-lasting."

Tebua Tarawa and Abanuea are just two small islands affected by rising sea levels in the South Pacific. The South Pacific Regional Environment Programme

(SPREP)—an organization comprising 21 island countries in the South Pacific among others—has expressed concern for not only the loss of Kiribati's two small islands, but also for severe erosion on the 29 atolls of the Marshall Islands, and the 200 inhabited islands of the Maldives where one third of their beaches are being swept away by rising seas.[45] Recently though, of an utmost concern to the SPREP is a tragedy befallen the people on another South Pacific island: Tegua.

Tegua is another one of those tiny islands in the South Pacific that, although sitting barely above sea level, has remained unchanged for thousands of years. It's not exactly the biggest island of the Torres Islands chain in the Republic of Vanuatu, but it does house a small population of indigenous people that have lived along its coasts for generations. The lives of Tegua's inhabitants are changing, in similar ways as the Inuits of the far north. Although the Pacific Islanders aren't concerned about polar bear or walrus hunting, and Teguans likely have no idea what sea ice or tundra are, their tiny island is starting to disappear. A coral reef that once served as a barrier to Tegua's fragile shorelines has been battered so hard in recent times that waves now breech the reef and collide with the island's coastline. This is causing erosion to accelerate between three to nine feet a year on the shores of Tegua, permitting pounding surf to wash away homes.

As a result, in late 2005 more than 100 people on Tegua had no choice but to relocate their homes to higher ground. But there were other problems caused by the rising seas on Tegua. Although the indigenous people of Tegua could be relocated to higher ground, drinking water on the island is scarce. Most of the freshwater on Tegua comes from springs that lay precariously close the coast that are revealed at low tide. Inland, there just isn't as much water. Now, with higher than normal sea levels around the island, the freshwater springs are either inaccessible, or contaminated with salt water.

The United Nations Environment Programme (UNEP) stepped up to the plate to help the people of Tegua in their relocation efforts. A project funded by the UNEP called Capacity Building for the Development of Adaptation in Pacific Island Countries, with additional funding provided by the Canadian government, helped to relocate the Teguan people inland to higher ground, and set up six 6,000-liter water tanks to harvest rainwater for drinking. If it not for this foreign assistance, the people of Tegua, accustomed to their way of life for centuries, may very well have been up the proverbial creek without a paddle. Slow changes over time may have allowed the people of Tegua to more easily adapt to a transforming environment, but the rising sea levels that washed away their homes and ruined their drinking water occurred over a very short period. Thanks to the intervention of the UNEP, the people of Tegua now have some hope for their future—at least for the time being.

The UNEP program that is helping the Teguan people has been set up to assist other low-lying communities endangered by rising sea levels, which seems to be an ongoing battle as of late.[46] As with the ICC and ANCAP organizations assisting Inuits in the battle of climate change, the UNEP has taken the stance that the climate change transformations that are causing the rise in sea levels and affecting the way of life for many around the globe are "the result of the rise in human-made emissions in the atmosphere." Unlike the ICC and ANCAP however, the UNEP doesn't appear interested in filing any petitions at the moment. Instead, they are working with other organizations like the Global Environment Facility (GEF) to develop programs to assist areas at risk of rising seas, and the effects of climate change. After all, other areas besides just Tegua are being affected.

One such area suffering from the incessant battering of rising seas is the Carteret Islands off of Papua New Guinea. There, some 2,000 people had no choice but to make relocation plans in late 2005 to the island of Bougainville where they will start their lives over again.[47] Ask any one of the people moving from the Carteret Islands, the relocated Teguans, the people of Kiribati, or even the Inuits of the north if global warming is real, and their answer will likely be, Yes. After all, they're seeing the effects of a warming world first hand. After living in their native lands for countless generations, a sudden change seems to be sweeping over their homeland, and subsuming it into the sea. But why?

There are three possible explanations as to why islands could disappear or shrink: (a) the islands are sinking (b) sea levels are rising or (c) sea levels are rising *and* islands are sinking. Like in most arguments, the debate on global warming tends to dwell on *either-or* situations, where only one of two possible answers could explain what is happening. Yet, there is always the third yet often overlooked alternative in any argument, which points to the fact that both parties could be right—at least to some extent. Such is the case that solves the mystery of disappearing islands in the South Pacific: islands (actually atolls) are indeed sinking; yet sea levels are undoubtedly rising.

Many atolls in the South Pacific are sinking—atolls are notorious for that. In fact, atolls are constantly sinking; that's how they're formed to begin with. Atoll formation theory tells us that at one time, a volcano rose high above the surface of the water, and when it became dormant or died, it began to settle and sink. Coral then formed around the fringes of this sinking volcano, thus making the reefs that surround each of these tiny islands. Sitting above old volcanic activity makes an atoll's foundation a bit unstable. When you consider that an atoll chain like the Carterets sits only a few feet above sea level, any shifts in tectonic plates below the ocean floor could certainly affect the placement of the tiny islet above. So it is entirely plausible that much of what is going on with the Carterets, Tebua Tarawa, Abanuea and Tegua could be from the natural process of atoll sinking.

But once again, this is not the entire answer to explain away the fate of the inhabitants on these tiny South Pacific islands.

The science of global warming is peppered with myriad phrases of ambiguity like "studies suggest," "scenario," "if," or "possibly." While much of these hazy dictions relate to the root cause possibly being linked to human deeds, when it comes to rising sea levels, descriptions are hardly obscure. No matter what may be causing it, sea levels have risen; this we do know. In fact, tide measurements around the affected Carteret atolls show that the sea levels have risen.[48] But then again, this isn't the first time in history that this has happened.

Earth has endured numerous ice age events, many of which have lasted over 100,000 years. In between these ice ages, the Earth goes through a warming period. It's a cyclic kind of thing; our planet gets hot, then it gets cold, reasons for which I'll explain in more detail in chapters 4 and 5. In any case, our last major ice age started up some 120,000 years ago and ended about 11,500 years ago. Before this ice age started, the warm period that preceded it saw sea levels over 15 feet higher than they are today. Part of the reason for higher sea levels in a warm period is that water expands when it is heated. Known as *thermal expansion,* the amount that water will grow in size varies depending on the temperature of the water before it is heated. For instance, at equatorial latitudes where seawater temperatures may average 25°C (77°F), an increase of 1°C could cause the water depth to increase by about 1.2 inches. Although an inch or two doesn't seem like a lot, observations do show that during the 20th century the average sea level around the globe rose anywhere from 4 inches to 8 inches, most of which was contributed to thermal expansion from warming waters.[49]

Melting ice though is another big contributor to increased sea levels, especially just recently. As more water melts from glaciers and ice sheets sitting up at high elevations, the meltwater runs downhill and eventually dumps out into the Earth's oceans. During our last ice age, glaciers and polar ice caps were really quite dense. So dense in fact, that when the ice age peaked some 18,000 years ago, glaciers and ice sheets had pulled so much water out of the world's oceans that the Earth's sea level dropped an astonishing 300 feet lower than it is today.[50] That's enough of a drop in sea level to allow Britain to be joined with the continent of Europe. Our planet was a much different place back then, and oceans were quite different as well. Now that we're out of that ice age, water levels have come back from the melting of glaciers and ice sheets. Since our last ice age ended around 11,500 years ago, a lot of water has had the chance to melt and fill up the oceans once again. Yet not much really changed for many years, and sea levels remained quite constant throughout much of our history. That however, has changed recently, and scientists are figuring out why.

Studies conducted in 2003 and 2005 by NASA and others show that in some cases, glaciers are melting faster than anyone had realized.[51] We may often think of ice as a solid that melts when it's heated, which is true, but glaciers are a bit different. Although they do indeed melt, glaciers are actually large, long-lasting rivers of ice. Just like a river of water, glaciers are always flowing. You'd need to be quite patient though to see a glacier flowing, since they move no faster than a lumbering pace of just a few miles per year. The NASA studies tracked the speed of moving glaciers covering Greenland, and found that something unique and alarming was going on—something directly linked to the recently rapid rising sea levels, and the fate of islands in the South Pacific.

In both NASA studies conducted in 2003 and 2005, researchers used satellite and other data to observe large changes in both speeds and thickness in Greenland's glaciers. In the first study, the comparison was done between 1985 and 2003, and the second study between 1996 and 2005. In the first study, the data showed that Greenland's glaciers slowed down from a velocity of about 4 miles per year in 1985 to around 3.5 miles per year in 1992, and remained somewhat constant until 1997. By 2000 however, the glaciers had sped up to a speed of nearly 6 miles per year, topping out with the last measurement in spring 2003 at nearly 8 miles per year. The study conducted in 2005 showed that speeds had reached nearly 9 miles per year.

Greenland glaciers have been moving at over twice the rate that they had only ten years prior. This in turn is causing a more rapid melting of the glacier ice, thus dumping massive quantities of ice and water into the sea. In fact, in 2005, 54 cubic miles of ice melted off of Greenland's glaciers, more than double of what was lost in 1996 when only 22 cubic miles of ice were measured as being lost.

Glaciers though are melting not only in Greenland, but worldwide as well. Coverage of the 2006 winter Olympic Games in Turin, Italy, showed how ski resorts in the Swiss and Italian Alps are desperately trying to preserve their melting glaciers by wrapping them in sun-protective plastic and fleece.[52] It's a monumental feat to say the least, with football field-sized sheets of white material costing around $70,000 each being dragged out by heavy machinery to wrap entire glaciers and protect them from warmer than normal temperatures. This new method of glacier sun protection has been going on over the last few years after ski resort owners became alarmed by observations showing that over the past fifteen years Swiss glaciers have lost about 20% of their surface area, and continue to melt away.[53]

Glacier melting is alarming, in that substantial stores of water are locked up in these colossal rivers of ice. Most of the world's frozen glacial water is held in Antarctica and Greenland, yet there are numerous other glaciers around our planet, like those in the Rocky Mountains, Alps, and Andes. If all the glaciers outside of Antarctica and Greenland were to melt, it's estimated that sea levels would rise nearly 20 inches.[54] But this isn't the only ice we need worry about. Polar ice is immense, and it too is starting to thaw.

A Canadian-led research project that began in 2002 and ended in late 2005 sampled the winter and spring conditions in the Arctic, and found that polar ice is melting at a rate of about 28,500 square miles each year, which is an area about the size of Lake Superior.[55] More astonishing though, is that this melting seems to have been going on over the last 30 years—that's a lot of lost ice, and a lot of resultant meltwater to raise sea levels, and combined with thermal expansion, deluge tiny low-lying islands in the South Pacific.

If you were to ask any Inuit of the north if Arctic ice were melting, they'd inevitably tell you "of course." It would be nothing short of obvious to anyone now living in the tundra-like territories where sea ice has either thinned or disappeared. And if you queried Pacific Islanders about this issue, they'd more than likely side with the Inuits as well. These two regions—the Arctic and low-lying islands of the South Pacific—have been seeing effects of climate change first hand, but they're not the only ones to observe the devastation from our growing oceans; rising seas from a now warmer world are affecting the United States as well.

THE FLORIDA PHENOMENON

The Sunshine State has been an increasingly popular place to live in recent years. With its warm weather averaging 68°F in the southern portions of the state in the winter and 83°F in the summer, 1,800 miles of coastline, 1,200 miles of sandy beaches, over 1,250 golf courses and no state income tax, Florida is quite attractive to those tired of shoveling snow out of their driveways in the northern states.[56] Real estate is far more affordable than other sunny beachside states in the U.S. with Florida's 2004 median housing cost at a mere $149,000 compared to California's whomping $391,000 median housing cost[57], thus further enhancing the appeal of this peninsula panhandle state. In fact, Florida has become so attractive over the past few years that between 2000 and 2004, the state's population grew from 15.9 million people to around 17.4 million, for an increase of 8.8%. During this same time, California grew by only 6% and the entire U.S. population barely grew by 4.3%.[58] And when looking to the chillier north, the state of New York grew a scant 1.3% during this time—obviously, people like it warm, affordable, and state tax-free.

There's no doubt that Florida's population is growing, and has been for quite some time. Although Florida has plenty of room for its new inhabitants, like most states that share at least one of their borders with an ocean, the majority of Floridians live in coastal communities. In fact, around 95% of all Florida residents live along the state's coastlines.[59] This is prime real estate, since the sandy beaches, warm water, and scenic beauty of palm tree-lined boulevards attracts people to Florida far better than swampy everglades or alligator-laden, mosquito-infested inland waterways—although they do indeed have a beauty all their own. Nevertheless, the water's edge not only draws people to its scenic shores for pleasure and pastimes; its allure is innate. Humans since the dawn of time have always flocked to riverbanks, lakeshores, and oceanfront strands of sandy beaches since living close to the water provides sustenance in one form or another—there may be fish to catch, or water to drink or irrigate crops. The people on sinking atolls in the South Pacific lived close to their coasts for similar reasons, but of course, nature as of late has been forcing them to higher ground. Florida, although still growing with incoming residents seemingly not aware—or concerned—of fates that can befall a region of land sitting precariously to the sea, is now facing the same threats from climate change as the people of the Carterets, Tebua Tarawa, Abanuea and Tegua.

Sea levels are not just rising in the Arctic and South Pacific—they're rising everywhere. South Florida's sea level has risen about a foot since 1846, and it is still rising today at a rate that is nearly ten times faster than the average rate of sea level rise along the South Florida coast during the past 3,000 years.[60] If this rate of rising seas were to continue at this pace, the sea along the Southern Florida coasts could climb another three inches by 2025 and ten inches by 2100. But that's a best-case scenario based on the recent rate of rising sea levels without taking into account the fact that sea level increases lag behind global temperature increases—it takes a while for ice to melt and water to expand after global temperatures rise, kind of like waiting for water to boil on a hot stove, or ice cubes to melt in a glass of tea. Future warming trends that are predicted to occur over the next few decades—taking into account the lag in sea level rise to temperature increase—are expected to accelerate the rise in sea levels by magnitudes. In fact, by 2100, South Florida's sea level could very well be two to three feet higher than it was in 1990.[61] This would be enough of an increase to permanently cover much of the populated region of Miami where elevations hover around three feet above sea level. And of course, there are myriad other low-lying cities surrounding Florida's three-sided coastal periphery that would also be subsumed by the sea with a mere two or three feet increase in sea level.

Currently, no one along the Florida coast need worry about relocation to higher ground—at least not yet, or for a couple more years at least. But if sea levels rise as predicted, then the vast majority of Florida's population will be faced with some serious problems. Even now, rising sea levels are starting to erode Florida's coastline. As the erosion continues, and if sea levels rise a couple more feet over this century, then it wouldn't be prudent to leave beachside real-estate as a family trust, since it would inevitably lose its value when passed down to your great grandchildren, possibly even sooner. Real estate tends to lose its value once the sea reclaims the land, and over time, if trends continue, that's exactly what will happen in the heavily populated regions of Florida's coastal communities.

The rising sea levels are also a concern to the largest remaining subtropical wilderness in the continental United States: Florida's Everglade National Park. The Everglades, already stressed by pressures of human development in the region, are facing increased salinity from intruding seawater. Since The Everglades is a freshwater marsh, intruding seawater would poison the region, much in the way it did the drinking water on Tegua. This increased salinization of the water, and inevitably the surrounding soil, would harm the Everglade's vast array of wonderful wildlife like the Florida panther, the West Indian manatee, the snail kite, wood stork, bald eagle, American crocodile, and of course alligators as well.

Increased salinity from intruding seawater could also harm agriculture in Florida, contaminating the soil with unwanted salt from streams and aquifers that become mixed with flooding seawater. Florida leads the southeast in farm income, produces about 75% of all U.S. oranges and provides nearly 40% of the world's orange juice supply. If sea levels become high enough that streams, aquifers and other waterways become swamped in storm surge, seawater will inevitably leach out onto land. This salinization of the soil could cost Florida farmers dearly, in turn hurting the state's agricultural revenue stream.

Although some of Florida's tragedies are still out a bit in the cloudy crystal ball—waiting for sea levels to rise a bit more—Florida is seeing increased sea levels now. Moreover, with much of Florida's population living no more than about twenty feet above sea level, Florida is a state on the brink, with every increase in sea level inching all that more close to a dangerous situation.

Florida is also sitting smack dab in the center of hurricane alley, and is known to get hit from time to time by nasty storms like Andrew in 1992, Opal in 1995, Earl in 1998, Irene in 1999, and Charley, Frances and Jeanne in 2004, just to name a few. These were all some of the deadliest and costliest hurricanes ever recorded.[62] And then came along 2005, a year that broke numerous hurricane records, which I'll elaborate more on in chapter 3. Hurricanes have the attention of everyone living in Florida, probably more so than the worry of future rises in sea levels that could eventually wash away their homes. With sea surface temperatures higher than normal, hurricane formation has become more pronounced in the Atlantic (which I'll explain in more detail in chapter 5 as well). Combine this with an ever-increasing coastal population flocking to the shores of Florida to snatch up reasonably priced coastal real estate, and the problems associated with enhanced hurricane seasons are exacerbated, leaving Florida more vulnerable to loss of land, property, and worst of all, life.

Compounding the problem of coastal protection, salinization and storm damage vulnerability along Florida's coastline is what is known as *coral bleaching*, or a dieing-off of coral. Florida has had a booming bounty of coral-forming reefs around much of the peninsula, and just like those tiny disappearing islands in the South Pacific, the coral reefs around Florida serve as a first-defense protective barrier from incoming storms, rising seas, and the like. Without a good shield of coral reef armor surrounding Florida, storm wave energy could intrude with greater ferocity, so it's in the best interest of Florida to maintain its massive coral reef structure. But this is hard to do when water temperatures warm up.

Coral may look like nothing more than a crusty pile of sea stuff with plant life growing off of it. But coral is a living thing, existing as small sea anemone-like polyps that form into great colonies. As these corals grow, they turn into calcium-crusted skeletons. And when they die, future generations of coral grow on top and alongside of them. If something kills off living coral, then the reef it was forming will no longer grow, and the crusted remains will eventually deteriorate from natural causes like erosion and sea creatures that feed on the calcium-rich coral. Warming the waters around coral is one way of harming it and can cause it to die. As coral dies, it looses its color, thus the name, *bleaching*.

Coral bleaching is underway around much of the globe. Florida is especially feeling the impact as well with almost all of its elkhorn and staghorn corals completely destroyed. It's estimated that over 90% of these usually strong, thick, antler-like corals have been decimated since 1980, primarily from bleaching.[63] Amplifying the problem of lost coral though is the destruction from hurricanes, which everyone knows is a problem from time to time around Florida's coastline. If hurricanes are also intensified by warmer waters, they can wreak even more havoc on the reef structures around the Florida coast from excessive and incessantly pounding surf. Thus, climate change can take the coral reef structures into a death spiral of bleaching and intensified storm erosion.

Take away Florida's coral reefs, and you leave a coastline vulnerable to increased storm damage. But there are other short and long-term economic concerns as well. Tourism is big in Florida, and in 2001 it comprised $57 billion of the state's total $491 billion gross state product, or roughly 11%.[64] Besides Epcot Center and Disney World, Florida's coast ranks near the top in tourism popularity. It's estimated that the attraction of the coral reefs around the Florida Keys alone brings in around $1.6 billion of revenue annually.[65] If coral loses its appeal, then tourism dollars could decline. This problem could be compounded with the waning interests of travelers journeying to a place known to have increasing risks of hurricane damage and depreciation of natural waterways from salinization of intruding seawaters. If so, Florida could someday see a further decline in tourist attraction and subsequent revenue.

Such is the phenomenon of Florida: its attractiveness may one day be its undoing. As more people flock to the sunshine state, climate change continues to unfold—sea levels rise, coastlines erode, salinization threats increase, coral reefs bleach and deplete, and hurricane destruction becomes a greater threat from not only warming waters, but also a growing population, placing more people at risk of a disaster. Still, Florida has yet to lose its luster, and while it sits on the brink of climate change impact, its residents and tourists continue to come, flourish and stay—for now.

A Bay to Remember

On December 20, 1606, Captain John Smith left his home in England to set sail for the New World on a voyage to colonize what we now know as Virginia for King James I. His voyage was filled with adventure and danger, facing harsh, brutal weather while forming the city of Jamestown. Constantly, Smith and his troops came under attack by the Algonquian Indians, who eventually captured Captain Smith and took him as prisoner, destined for execution. Luckily, the chief's legendary daughter Pocahontas threw herself in front of Smith, begging that his life be spared. Fortunately for Smith, the Algonquian chief listened to his daughter's pleas and felt mercy. And so it was that this tale had a happy ending and Captain John Smith lived to see another day—many more in fact that led to new discoveries, including Chesapeake Bay, a region that's been drastically transformed since the days of Pocahontas.

In 1608, about one year after Captain John Smith's life was spared by Pocahontas' father, Smith and his crew sailed up Chesapeake Bay on the mid-Atlantic coast, eying scenery far different from what we see there today. Smith and his crew passed a large, low-lying island in the upper middle reaches of the bay, about 34 miles south of today's Baltimore. Poplar Island, as it became known, covered more than 1,400 acres back in those days nearly 400 years ago. This island established a population and eventually supported a small town, and later—in the 1930s and 1940s—it became an exclusive retreat for politicians, including Presidents Franklin D. Roosevelt and Harry Truman. Even these two remarkable presidents saw Poplar Island differently than any observer will today.

You'd have to look pretty darn hard to find Poplar Island in the Chesapeake Bay today. Although 400 years ago this 1,400-acre land mass was an easily visible landmark in the Bay, Poplar Island today consists only of a small group of islets that together measure less than five measly acres. Dead trees rise from the water to mark where land once stood only a few years before. Rising seas and heavy erosion have flooded and eaten away more than 99% of Poplar Island during just the past 150 years. With the shoreline retreating by more than 13 feet annually, the island's remnants would have completely vanished under the sea by the year 2000 if it not for a $427 million restoration program launched in 1998 by a coalition of federal, state, and nongovernmental agencies to rebuild this disappearing, historic island.[66]

Poplar is not the only island dipping below the horizon in Chesapeake Bay though; it's just one of at least 13 islands in the Chesapeake Bay that have disappeared entirely since the region was first described and mapped by Europeans back in the 1600s. Smith Island (named after the legendary Captain) located near the mouth of the Potomac, supports three small towns and a population of close to 400 people, but since 1850, this once heavily wooded island has lost 30% of its land, and has split into several marshy islands. Tangier Island is also vulnerable and in peril, threatening the way of life of the watermen and crab pickers that inhabit this Chesapeake Bay island.

The islands of the Chesapeake Bay are seeing similar problems as the tiny atolls in the South Pacific: sea levels are rising, yet land is also sinking. In the case of Chesapeake Bay however, one factor is exacerbating the other: rising seas are causing land to sink ever more quickly, throwing the islands of this region into a downward spiral—literally.

Tide gauges around the Chesapeake Bay indicate that the sea level in the Bay is rising at twice the average global rate.[67] While the average rate of sea level rise around the planet has increased by 1-2 millimeters each year during the 20th century, Chesapeake Bay's sea level has increased at an alarming rate of 3.4 mm per year. Since sea levels around the globe are on the rise, we know that some of the islands around the Bay are disappearing as water levels climb. We've seen this elsewhere in the South Pacific, Florida, and the Arctic as well. But with the level of sea level ascent doubling, there must be something else afoot.

Marshy areas like the Chesapeake Bay region are recuperating from what is known as a *post-glacial rebound*, where once frozen land continues to settle after it's warmed up enough to melt. This is evident in some regions of Alaska's interior where thawed permafrost is now causing land to sink by as much as 33 feet in some areas. These areas of the Arctic north have only recently thawed, which can explain their recent and sudden drop in land elevation. Chesapeake Bay though thawed out a long time ago after our last ice age ended some 11,500 years ago; it's had plenty of time to settle, but it's not all settled just yet. The extremely moist ground supporting the Bay's islands is still acclimating. This is natural, and expected. What is irregular however is a foreign, influential factor of climate change that is aggravating the sagging.

If sea levels gradually increase, marshy land, such as that around the Chesapeake Bay can keep pace with the rising water levels by trapping sediments and their own organic silt. This raises the bottom of the marshy land, thus offsetting the rise in water level. It's just another one of nature's miraculous ways of maintaining equilibrium. Mother Nature is good at that, as long as changes occur gradually. If however, the sea level rises significantly faster than the rate at which the marsh can respond, the marsh will continue to sink, drown and become lost. Such is the case with the islands in the Chesapeake Bay; rapidly rising seas are outpacing the marshland's ability to replenish itself, so its natural subsidence is intensified, kind of like struggling while stuck in quicksand—you only speed up the sinking process.

Besides the concern for loss of land and human habitat, and the beauty of this historic and majestic land enjoyed today by thousands of kayakers, anglers, windsurfers and birders, there is also the matter of impacts on local wildlife. As shorelines are lost, so are habitats for myriad species of migratory waterfowl and fish. American black ducks, colonial water birds, ospreys, eagles and various migratory birds thrive on the isolated upland habitat that's within close proximity to the water's edge. Most of this habitat is found on the Bay's disappearing islands, which has been greatly reduced over the years.[68]

As waters warm, oyster disease becomes a concern as well, being able to spread more easily to infect the surrounding fish populations. Parasitic diseases such as MSX and Dermo have already taken a toll on Chesapeake Bay's oyster population, brought on by poor water quality and harvest pressure. Yet these diseases are easily intensified and become quite prolific when waters warm by only a degree or two and become saltier as well. Tragically, the native oyster population in Chesapeake Bay has dwindled to only 1% of its historic level, which has had a dramatic economic impact on the oyster industry in the area.[69]

Chesapeake Bay is an attractive place to live, with a population growth of 28% from 1970 to 1997. With today's population around the bay near 15 million, 300 new people each day move into the area to call this historic land their home. It's estimated that by 2020 the population around the Chesapeake Bay region will grow to over 18 million.[70] As construction on homes and infrastructure continues, over time this region could become further stressed.

Chesapeake Bay is another coastal locale on our planet that sits in harm's way of rising seas and impacts of a warming world. Climate change puts regions like the Chesapeake Bay and other coastal areas in the danger zone. But shorelines are not the only geographic terrain of concern; some areas in fact aren't getting wet enough.

DUST IN THE WIND

Reese Woodling remembers mornings when he'd walk along the grounds of his Arizona ranch and come back with his clothes soaked by morning dew from brushing along the tall, moist grasses on his land. But by 1993, his morning walks became arid. When rain stopped falling in 1990 and a drought in the American Southwest thus ensued, dew that was once prolific around his 2,700-acre ranch in the Malpai borderlands straddling New Mexico and Arizona became nonexistent. Instead of lush green grasses, rich from moisture spreading far and wide across his once fertile land, shrubs were taking over and infiltrating his ranch. Five years later, after nearly all the grasses dried up and subsequent food ran out for his cattle, Woodling ended up selling off half his cows. By 2004, he abandoned the ranch altogether.[71]

The American Southwest has been in the grips of a torrid drought for over a decade with Texas—one of the hardest hit states from the bone-drying drought—reporting a loss of $1.5 billion from the unusually parched landscape.[72] Other states in the Southwest are feeling the loss of agricultural revenue as well. And when hearing of other droughts around a warming world, the finger of blame turns once again to global warming.

This latest drought to hit America's Southwest states is not the first dry spell to pass through the region though. America has seen some extremely devastating droughts throughout the years including the historic "dust bowl" in the 1930s. And the most severe drought to ever hit Arizona occurred about thirty years ago between 1973 and 1977.[73] But the drought affecting America's Southwest today is a doosey.

According to the United States Department of Agriculture, a good deal of Arizona, Texas and Oklahoma are in the second highest rated drought intensity known as a D3—the highest is D4, which is affecting parts of Northern Texas and Southern Oklahoma.[74] The period between September 2005 and January 2006 had the driest months ever recorded in Arizona in well over 100 years.[75] Wildfires that began in the early winter of 2005 in the Southern Plains continued into January of 2006, with more than 330,000 acres being consumed by the end of January 2006, and Phoenix, Arizona set a new, all time record around this time: 104 consecutive days without rain.[76] So while Inuits continually worry about melting ice and coastal regions around the globe grapple with rising seas, there are some regions like the American Southwest that are just begging for rain as their climate has changed, and for the worst.

"Climate change" does just that: it changes climate. While some areas can get soaked, others suffer from extremely dry conditions. Some of this will vary from year to year or decade to decade from phenomena like El Niño and decadal oscillations that I'll cover in more detail in the next chapter. Yet areas that are prone to dryness become even drier as the air heats up and evaporation increases. This can literally suck any remaining moisture out of the surrounding landscape and water supplies, thus worsening an already arid environment. So when things heat up, drought prone regions become evermore susceptible as an already desiccated region quickly loses what precious water it has left.

So once again, we face a situation that cannot be answered by *either-or* responses, but instead, both sides of the argument in this case have truths to them: (a) the U.S. entered a natural drought and (b) temperatures are increasing. The combination of both issues is what's making the U.S.'s current drought all that much worse, and ringing the global warming alarm bell. If it were not for the increasing temperatures of late, the rate of drought and local evaporation would likely slow, and thus have a better chance at recuperation. But things are heating up.

The year 2005 was the hottest year on record, with temperatures likely not this warm for over 10,000 years.[77] The global average surface temperature tipped the mercury at 58.3°F in 2005; that's 1.4°F warmer than it was 100 years ago. A recent study also shows that the warming trend we're in the midst of is the longest recorded in the past 1,200 years.[78] Many global warming skeptics contend that the well-known "Medieval Warm Period" that lasted from the 10th century to around the 14th century is proof that our planet goes through abnormal warming trends from time to time. Yet this recent study also shows that while the Medieval Warm Period affected many regions of the planet, the most *widespread* warmth was found not in the Middle Ages, but during the 20th century. This 1,200-year warming study, conducted by Timothy Osborn and Keith Briffa from the University of East Anglia England and published in the journal *Science*, is recent news in early 2006, and now seems to dispute the longstanding argument that our current droughts, rising waters, and other apparent effects of global warming may just be natural.

As global temperatures climb to new heights, more and more people are feeling the heat in drought prone regions of our world. In 2005, severe droughts continued in Eastern Ethiopia, Southern Somalia, portions of Tanzania and Northern and Eastern Kenya. The president of Kenya declared a state of national disaster in areas affected by severe shortages of food and water resulting from the prolonged drought. In Burundi, at least $75 million of the government's $168 million emergency funds for 2006 were earmarked to feed the country's drought-affected population, and an estimated 11 million people were faced with food shortages throughout East Africa and the Horn of Africa due to the ongoing drought in those regions.[79]

As Africa dries up, it blows away—literally. Two to three billion tons of dust is blown around the world annually, much it coming from Saharan Africa where dust storms have increased ten-fold over the past 50 years.[80] This dust usually travels westward and eventually crosses the Atlantic. Africa's dust has traveled so far at times that it's been detected as far west as Cape Cod.[81]

These more frequent African dust storms, brought on from the extended drought and intensified from warmer than normal temperatures, cross through regions like Israel before making their way to the Atlantic. Israel used to see these dust storms during the drier spring and summer months. But now, they're seeing these storms all year round.[82]

As Africa's dust continues its westward journey and travels across the Atlantic Ocean, some of the dust drops into the water, which can be a hazard to coral already under threat from warming waters. The Saharan dust carries a fungus known as *Aspergillus sydowii*, which is known to fall into the Caribbean Sea. This fungus is thought to damage coral by causing infections like sea fan disease and black band disease, which can lead to coral bleaching—a problem already under-way around much of the world's coral reefs from recent climate changes. While the study of African dust impacts on coral death is still under investigation, the U.S. Geological Survey has made a link to the increased frequency of dust storms in Africa, and dust records in the Western Atlantic.[83] While it can be argued that the dust contains nutrients and has even been known to help feed plankton[84], which in turn can then intake more carbon dioxide, which in turn should help diminish human-induced greenhouse gas concentrations, it's the nasty fungus being carried from Saharan Africa that is of concern. Couple this with a now increased rate of dust storm occurrences and the existing problem of coral deple-tion, and the trickle-down effect it has on coastal regions like Florida multiplies by magnitudes.

Not all fungus that grows in the sand is harmful however. In fact, some of it is actually quite tasty. Fat choy, a black, hair-like moss also known as *Hanyu Pinyin*, is essentially a type of freshwater algae that has adapted and grows well in the Gobi Desert. Chinese adore fat choy, and consider it a delicacy, using it in espe-cially large quantities in meals on the Chinese New Year. But harvesting the stringy black desert fungus takes a devastating toll on the expanding deserts of Northern China, which are also feeling the heat from climate change. To round up just one small bag of the deep-rooted fat choy, nearly 25 acres of desert need to be ripped up and displaced, thus ruining the land for future growth.[85]

China has mounted various efforts to try and curb the continued desertification of their northern regions. Today, as much as 900 square miles of farmland in Northern China is blown away each year.[86] That's twice the size of Hong Kong. Evidence of this immense loss of land to the whim of the wind has been seen in the U.S. when in early 2000, an unusually large dust cloud that originated in Northwest China drifted across the continental U.S. and parked itself over Denver, at times obscuring views of the Rocky Mountains. Besides the fact that this dust could contain coral-killing fungi, other fears of pollution are in the air as this dust can contain myriad other fine-grained particulate matter that can irritate lungs, complicate asthma, and cause other upper respiratory ills.

Animal diseases could also be possibly transmitted by dust storms, such as the dreaded foot-and-mouth disease that was reportedly spread to livestock in the U.K. from Saharan Dust in 2001.[87] If drought-stricken regions of our planet are continually affected by a warming world, the risk of proliferating pathogens becomes paramount—not just for humans, but our livestock as well. But dust from drought intensification isn't the only culprit that could increase the risk of disease across our planet. In fact, many regions around the globe are experiencing what happens when bacteria and viruses live in a warming world.

PARASITIC PLANET

When it gets warm, and if things especially get moist, bacteria and viruses live in a virtual Club Med. The tropical regions of our planet have long been known as breeding grounds for such infectious diseases like malaria, dengue fever, viral encephalitis, and yellow fever. These devastating diseases are typically limited to climates conducive to their breeding that usually requires at least 60°F temperatures. Thus, low-lying jungles that well exceed these temperatures serve as excellent breeding grounds for these viruses. But in a world that's warming, their range is spreading to new heights.

In Mexico, dengue fever has now spread above its former elevation limit of 3,300 feet, and has appeared at elevations as high as 5,600 feet. In Central America, dengue fever spread above its former limit of 3,300 feet and has been reported above 4,000 feet.[88] A recent assessment in 2005 by the World Health

Organization shows that diseases such as dengue are spreading, and killing more people due to our recently warming climate. This report, which was later published in the journal *Nature*, shows that there is indeed a direct link between "human-induced changes in the Earth's climate" and at least 5 million cases of illness and 150,000 deaths each year.[89]

As summers last longer, and winters shorten, people are now dealing with ticks and mosquitoes for longer periods than usual. There is evidence that warmer temperatures are spreading tick-borne encephalitis in Europe, which can increase the risk of malaria.[90] These longer warm seasons are also allowing pollinating plants to proliferate, thus causing more distress for allergy sufferers. Nevertheless, some sneezing and doping oneself with antihistamines is a far cry from getting a serious disease from a blood-sucking mosquito, and that is of grave concern.

West Nile virus, which originated in Uganda in the 1930s, stayed fairly well isolated in the warm climates of equatorial Africa. Only recently has this virus been able to make its way to the U.S., introduced a short time ago from mosquitoes that can now expand their range with more warm-climate regions to buzz about in. Although birds were likely to have carried a West Nile infected mosquito to the U.S., warm weather has allowed the disease, and the mosquitoes carrying it to thrive. According to the Centers for Disease Control and Prevention (CDC), 2005 saw an astonishing 2,949 cases of West Nile virus with 116 deaths. In 2000, there were only 21 cases, and 2 deaths.[91]

As with dust-borne viruses and fungi, insect-borne diseases harm not just human populations but livestock as well. Japan is now dealing with foreign, African-originating viruses like Shamonda, Akabane, Peaton, Sathuperi, and D'Aguilar in recent years.[92] These dangerous mosquito-transmitted viruses are known to cause birth abnormalities and miscarriages among the cattle in Japan, a problem they've only had to deal with in the last few years as climate temperatures rose, and disease carrying mosquitoes became more abundant.

Yet, even though rising temperatures are assisting viruses and disease-carrying insects to flourish, there is possibly some good news out of all of this. Since winters are being shortened, it is thought that upper respiratory ills may be quelled, since the life span of wintertime viruses would be diminished in a shortened winter. While this is yet to be seen from a scientific perspective, one report issued earlier this year in London, did make a link to reduced cases of the respiratory syncytial virus (RSV).[93] This serious upper respiratory ailment has seen a shorter season over the past two decades, which researchers have attributed to the shortened (and less severe) winter months that Britain has recently experienced. Nevertheless, for now, there seems to be a much greater concern for other viruses that are seeping out of the equatorial regions and migrating northward, thus expanding their range and resultant devastation.

Our planet is a pretty darn big place. The Inuits, Teguans, Floridians, ranch hands of the American Southwest and the people from other regions mentioned in this chapter, represent just a few of the lives of the people on Earth being transformed by our changing climate. People of Arctic regions are watching their hunting grounds (and ice) disappear, Pacific Islanders are seeking higher ground, residents of the Chesapeake Bay are now very much concerned about lost islands, and people from all walks of life in drought stricken areas of the world are grappling with ways to feed themselves, and eek out a living. Times, they are a changing, and so is the climate that is causing the alterations of human life around the globe.

Be it from the deeds of humankind or not, our planet is on the move, taking on new shape in a rapidly changing state of climate flux. Interestingly enough though, all it took to spin the wheels of change was a few small, seemingly insignificant tweaks to our world's environment.

3

The Goldilocks Complex

Science is organized common sense where many a beautiful theory was killed by an ugly fact.

—*Thomas H. Huxley*

Science can sure be a funny thing at times, particularly with the theories and correlations that make scientific research such a mystery. Take for instance Einstein's infamous 100-year old formula, $E=mc^2$, which partially describes his *special theory of relativity*. Without this theorem, the atomic bomb would never have been made, we'd have never landed a rover on Mars, and the discovery of Black Holes in space would likely not have ever happened either. Yet this well-proven formula, that's so highly ubiquitous today in the world of science, math, and engineering is still called a *theory*, and not a fact or law.

Like all sciences, the study of global warming also has its collection of theories, some of which are based on correlations where instances of one thing are tied to instances of something else. For example, there is a correlation between increases in global temperatures over time and the increase in carbon dioxide emissions into the atmosphere over that same period—specifically since the Industrial Revolution. This is one correlation that ties two instances of something together, yet other correlations are needed to more properly tie cause to an effect, thus showing more than one exhibition of proof. To do this, science often uses an overlapping correlation. In climate change science, an overlapping correlation of carbon dioxide levels in the atmosphere is linked to something else, say, a rise in sea levels. We now have two correlations: (1) a correlation between increased global temperatures and increased carbon dioxide emissions, and (2) yet another correlation between increased carbon dioxide emissions and rising sea levels.

These two overlapping correlations can be used to hypothesize that all three of these things—increased temps, increased carbon dioxide, and increased sea levels—are all directly linked; thus, allowing climate change theory to state that (a) increased carbon dioxide emissions are linked to increasing temperatures and (b) increased carbon dioxide emissions are linked to rising sea levels. Thus, climate change theory can deduce that (c) increased carbon dioxide levels are linked to rising temps and sea levels. Don't worry, there isn't a test at the end of the chapter, so you won't be graded on correlation linking. The key to this though, is that many elements of discovery have to be commonly linked to each other to come up with a theory, which can later be studied and proven to be true or false.

In the judicial system, this kind of theorizing by means of correlations is referred to as *circumstantial evidence*. People have been acquitted and put to death on these correlations alone. In well-proven theories like Einstein's special relativity though, many physical experiments have been conducted to prove out the ideas and formulas, which gains more supporters than just a theory with no physical evidence. Nevertheless, circumstantial evidence consisting of correlations can be quite damning, and if researched properly, these theories can be proven to be true. But correlations can be tricky and oftentimes deceiving.

For instance, one could deduce a correlation between storks landing on housetops in Florida to the birthrate of Floridian children, and then compare this to stork landings on housetops and birthrates in say, Georgia. It may be that during some sample of time, Florida had a higher human birthrate than Georgia, and it could also be that during this same time storks seemingly favored roosting on Florida housetops more than they did in Georgia. This correlation would lead to the theory that there is a direct link between storks roosting on housetops to human birthrates: the more often storks land on a roof, the more likely it is for women to conceive. But is there actually a link here? Probably not. Through further scientific research, the stork to birth correlation could easily be proven to be nothing more than an old wives tale. This theory wouldn't stand up for long. The correlations comprising the science of climate change though are a bit more elusive.

Mother Nature is a tricky ole gal, and just when we think we know everything about her, she throws a curve ball our way. One thing we have learned from observing our world's climate and all of its correlations, is that Mother Nature has some serious issues, including a Goldilocks Complex. Just as Goldilocks likes the temperature of her porridge just right—not too hot, yet not too cold—so does Mother Nature in her desire for air and sea temperatures around her world to maintain a steady state of equilibrium. If we disturb the balance of nature by even the smallest of perturbations, butterfly-effect-like catastrophes can occur. These seemingly small disturbances however provide even more insight into the correlations that help us to understand the world we live in, and if global warming is due to the deeds of humankind or not. With the elements of nature so tightly coupled that provoking one feature will cause agitations elsewhere, Mother Nature's Goldilocks Complex provides us with countless correlations.

In chapters 4 and 5, I'll be weighing in both sides of the global warming debate, showing first in chapter 4 the view from skeptics that Mother Nature is just doing her thing and all is well, and then the view from believers in chapter 5 that humans are to blame for our warming world. Many of the arguments surrounding the global warming debate deal with theories and correlations. Yet as I've mentioned before on the topic of answers to a problem, the outcome may not be a binary result of *either-or*—just answer A or answer B. The complex world in which we live rarely allows explanations to come so easily. Yet using Mother Nature's Goldilocks Complex to our advantage, we can often tie answers A and B together with surprising correlations.

To assist in finding and binding answers to reasonable correlations, we can turn to some of Mother Nature's most unique phenomena, such as recent hurricane intensification, the El Niños of recent times, ocean oscillations, and other such things that, when lightly perturbed, can drastically set-off Mother Nature's finicky Goldilocks Complex. Understanding the effects of these climate-sensitive phenomena will help us to understand the causes to which they are correlated and linked.

THE TEMPEST'S RAGE

During the second week of October in 2005, only a couple weeks after Hurricane Rita slammed the Louisiana coast—a region still recuperating after the devastation from Hurricane Katrina—something strange was going on in the Atlantic that caught the attention of meteorologists at the National Hurricane Center. An unusually large, monsoon-like circulation and a broad area of disturbed weather were developing over much of the entire Caribbean Sea, spanning far and wide. Mixing with additional disturbed activity known as tropical waves, over a period of a few days this behemoth storm continued to grow. By noon on October 15th, this storm, which would soon break records unimaginable, became cyclonic and acquired its birth status as an official tropical depression. Two days later, this storm was given a name: Wilma.

Although Katrina and Rita held the spotlight from their relentless devastation on the Gulf Coast, Wilma was much stronger, bigger, and ferocious—almost unearthly in fact. When measuring the power of any storm, scientists use an increment of pressure known as millibars. This is used to measure how much pressure there is in a high or low-pressure system. Normal pressure at sea level runs around 1013 millibars. Sunny weather associated with high-pressure registers greater, and foul weather (in low-pressure systems such as storms) measures much lower. Normally, hurricanes run around 950 millibars, which in itself is quite a bit lower than the normal 1013. Rita and Katrina though had impressive low-pressure readings measuring 897 and 902 millibars respectively. But with Wilma, a new Atlantic hurricane record was broken with an intense low pressure of 882 millibars. What made Wilma even more unique though is that her pressure dropped so suddenly, like nothing before. In one 24-hour period, this storm dropped an amazing 95 millibars, occasionally dropping 8 millibars per hour. Meteorologists, still on edge from lessons learned by lethal storms that preceded Wilma that year, struggled to keep up with this rapidly changing and potentially deadly, tumultuous tempest.[94]

Nothing can sound the climate change alarm louder than a devastating storm. Slow changes like rising sea levels or melting ice may leave us with a kind of landscape amnesia and creeping normalcy, but being pummeled by a hurricane can certainly be a wakeup call. In 2005, the Atlantic was rocked by record-breaking storms that will inevitably rewrite chapters in our world's climate history. With massive destruction and loss of life, many searching for answers have been looking to find a cause for the intensification and mind-boggling number of storms that swept through Florida and the Gulf Coast. Almost immediately, global warming came to the minds of many. But can any of the events in the 2005 hurricane season be linked to climate change? Perhaps yes, perhaps no. There are reports from 2005 that were published in the journals *Science* and *Nature* that discuss this in more detail, and I'll cover these more in chapter 5. What we do know however, is that conditions became prime for hurricanes that year; thus, we have one correlation that hurricanes did form and became stronger than normal by favorable hurricane-formation conditions. The link to bind cause to the effect however is a bit of a thornier issue, yet there are some interesting associations between the 2005 hurricane season and our changing global climate.

Record-breaking events tend to grab our attention, and for good reason. When meteorological records are broken, then either something is going on in our world that we've never seen before—perhaps a fluke of nature—or perhaps the cause for the hurricane intensification is indeed tied to climate change. Records provide excellent indicators that serve as encouragement to dig deeper for answers, and the overly active 2005 Atlantic hurricane season had plenty of new records to consider.

In all the history of the Atlantic, 2005 was indeed the busiest year on record with 27 named storms; the previous record was set in 1933 with 21 storms. Out of those 27 named storms, 15 became hurricanes, breaking the record of 12 back in

1969. Four hurricanes made landfall in 2005, breaking the record of three only the year before. And of all the hurricanes that did form, three of them gained the highest rank of category 5, breaking the record of only two category-5 storms in 1961. And as mentioned earlier, 2005 had the strongest ever-recorded Atlantic hurricane, Wilma, clocking in at 882 millibars, breaking the record made by Hurricane Gilbert in 1988, which had 888 millibars. Wilma came awfully darn close to breaking a world record as well, vying just slightly weaker than Super Typhoon Tip[95], which on October 12, 1979 clocked in at 870 millibars—Wilma had 172 mph winds, Tip 190. Nevertheless, Typhoon Tip formed in the Northwest Pacific Ocean, and Wilma in the Atlantic, thus making Wilma the strongest ever hurricane to form in the Atlantic (but not worldwide, yet close to it).

In addition to the strong Atlantic hurricanes that formed in 2005, that year also saw the strongest July hurricane—a time when things are barely rolling in hurricane alley—with Emily topping out at wind speeds of 155 mph. Her runner-up was Dennis, which formed only one week earlier with 150 mph winds, breaking the hurricane record in 1926 of a storm that blew in at 140 mph in July.[96] And just when we thought the hurricane season was done and over with in 2005, and we were all preparing for the new year, Zeta formed on December 30th—a month after the official end of the hurricane season—tying 1954's Alice as the latest-forming storm of the year. Yet Zeta went down in the record books as the first storm to ever survive so long into the month of January—Zeta finally fizzled out on January 6, 2006.

In 2005, for the first time since storms began acquiring names back in 1953, the year's allotment of storm names was exhausted, requiring the use of the Greek alphabet to name the season's last storms: Alpha, Beta, Gamma, Delta, Epsilon and that surprising late season storm Zeta. There is no doubt that 2005 was busy, and yet it was costly as well with an estimated $125 billion in damage (and climbing). Loss of life reached tragic levels as well, with Katrina being the deadliest Atlantic hurricane since 1928 with over 1,300 lives lost to that storm alone.

Although the debate continues over whether or not global warming is linked to 2005's devastating hurricane season, we have seen first hand Mother Nature's Goldilocks Complex in action. Small changes, similar in nature to the proverbial facile flapping of a butterfly's wings, set off a virtual maelstrom of violent storm activity. What's more interesting, is that in the case of the 2005 hurricane season, it was not just one single factor that perturbed our planet's normal cyclone season; instead, there were many things at play.

First, water temperatures were undoubtedly higher than normal with North Atlantic sea surface temperatures running about 1°C (about 1.8°F) above normal in 2005. In 2004—another very active hurricane year in the Atlantic—the water temperatures were slightly lower, but still above normal at around 0.5°C higher than they usually are. By comparison, in 2000, a year with only 8 hurricanes—about half as many as there were in 2005, and none of which made landfall in the U.S.—the sea surface temperatures were running about normal, or slightly below.[97] Thus, a link could be made to increased sea surface temperatures and the number of hurricanes. Still, this is only one correlation amidst the intricate workings of Mother Nature's Goldilocks Complex, and one that skeptics argue does not make sense in a warming world.[98]

The argument is that on a planet where everything is heating up, air above the warmer than normal ocean waters would serve as a kind of equilibrium to storm intensification. Hurricanes, like all storms, require instability in the atmosphere above them. First, the ocean water has to be 82°F or warmer for a hurricane to form. Second, the air high above the hurricane needs to be cooler. This allows the warm air, stirred up from the surface of the ocean by cyclonic winds from thunderstorms and other such smaller disturbances, to spin the storm into a tropical system kind of frenzy. The key in all of this, is that warm air from the surface of the water can rise more quickly, and whip a tropical storm into a massive hurricane if the atmospheric temperature above the hurricane is relatively cooler than the rising warm air from the ocean's surface. So the cooler the air in the upper atmosphere, and the warmer the waters on the surface of the ocean, the faster this air can rise, thus causing more cloud formation, and thus intensifying a hurricane all that much more.[99] Thus, the argument concludes that it's the disparity in temperature that makes all the difference, and if the whole planet were warming, then it would stand to reason that the air temperatures are warming in the same manner as ocean waters are; therefore, this should, in theory, equal everything out.

While global warming skeptics use this as a powerful argument, it serves more to cast a shadow of doubt on the link to warming waters and recent hurricane events, and is not well backed up by other explanations. This kind of defense worked well in the infamous O.J. Simpson trial, where only minimal—but not reasonable—doubt had to be shown, and no alternate explanation was required to acquit. Additionally, the global temperature increases in the past 20 years or so have been quite steady, yet as mentioned earlier, there have been extreme variations in sea surface temperatures in the Atlantic in just the past few years. While this is hardly a smoking gun, it is something worth considering. Nevertheless, even this could be linked to other phenomena, including the decadal oscillations I'll be covering later on in this chapter.

The second factor in the Atlantic's hurricane-related Goldilocks Complex that stoked the cyclone stove in 2004 and 2005 is what's known as the Loop Current. This warm, deep-water ocean current in the Gulf of Mexico comes up from between Cuba and the Yucatan peninsula, moves north into the Gulf of Mexico, and after swirling about for a bit in the Gulf, heads east and south through the

Florida Straits into the Atlantic. Katrina, Rita, and Wilma, once entering the Gulf of Mexico, encountered the Loop Current, which is thought to have intensified all three of these storms.

The sea surface temperatures in the Gulf of Mexico[100] were already quite high in 2005, but usually these warm temperatures are only at the surface, and not in deeper water below a hurricane. As a hurricane grows, it sucks up more and more water from below it. If this well of water below the hurricane is relatively cool, then it literally cools off the hurricane as it's sucked into the cyclone, and the hurricane then has a much harder time growing. If however warm water continues to be drawn into a hurricane, then the hurricane will be pumped with more fuel and become stronger. The Loop Current, being a flow of warm water far beneath the surface of the ocean, is just what a hurricane needs to grow. In 2004 and 2005, it's believed that the Loop Current was a bit stronger, but this mysterious ocean circulation continues to elude predictability, and science has yet to link its changes definitively to any particular phenomenon. Nevertheless, we do know that one of the influences placed on the deadly three of 2005—Katrina, Rita, and Wilma— was the warm waters of the Loop Current. All three of these systems passed over the warm depths of the Loop Current, and like a boost of steroids, the Loop Current pumped these hurricanes into ominously sized, savage, havoc-wreaking storms.

The third influential factor in recent hurricane seasons is the Bermuda High. This area of high-pressure sets up around Bermuda and stays put for much of the hurricane season in the Atlantic. High-pressure systems, as their name implies, have a lot of pressure to them, pushing air downward. This air, in turn, warms up and dries out. In this warmer, drier air, fewer clouds can form, and is why high-pressure systems are associated with clear blue skies and nice weather. But when the sky lacks clouds, the Sun can peer through all that much better, and bake the water on the ocean within and around this high. In the summer of 2005, the Bermuda High grew in size, and shifted a little more towards the southern portion of the U.S.[101] This slight shift and growth in the Bermuda High was a disaster waiting to happen.

With the Bermuda High located closer to the U.S. and being larger than normal, the Sun was now able to cook the already warmer than normal waters near the U.S. coastline, thus providing more fuel for forming hurricanes. But the growth and shift in the Bermuda High also set up a dangerous hurricane track. High-pressure systems in the northern hemisphere spin clockwise, and the Bermuda High is no different. As this massive high-pressure system spins, hurricanes passing under it get caught in the periphery of its circulation and get steered one way or another. If the Bermuda High is small, and located where it normally is, passing hurricanes caught in the spin of the Bermuda High are thrown northward, thus being diverted away from a direct hit on the east coast of the U.S. But in 2005 and in 2004, the Bermuda High was not sitting in its usual position. With its leading edge precariously close to the east coast, the Bermuda High was swinging incoming hurricanes directly at America's eastern seaboard and the Gulf of Mexico.

Assisting the Bermuda High in 2005 was a favorable African Easterly Jet (AEJ). The AEJ, originating over Western Africa, helps to steer hurricanes towards the Bermuda High, which can then spin the storms towards the U.S. In 2005, the AEJ was ideal for guiding storms along the Atlantic, the Bermuda High was poised to push along incoming storms to the U.S., and another weather factor, known as *vertical wind shear*, was not as prevalent as it normally would be.

Vertical wind shear can make or break a hurricane trying to form. In this condition, winds at higher elevations are blowing so much stronger than winds below. Rising hurricane columns are literally sheared off once encountering the stronger high elevation winds, thus killing the hurricane by lopping off its head. Unfortunately, 2005 and 2004 were years with reduced vertical wind shear—yet another one of the tiny elements of Mother Nature's Goldilocks Complex that was responsible for hurricane intensity and devastation.

The record-breaking 2005 Atlantic hurricane season had many variables at play, making it tough to point the finger of blame at just one cause, and thus gather direct correlations that could link the anomalous season to global climate change. The myriad elements behind the intense 2005 hurricane season created confusion on both sides of the global warming debate as most everyone was looking for one single answer to a rather complex issue. Warming waters, the Loop Current, an intense and abnormally positioned Bermuda High, a favorable jet stream, and reduced vertical wind shear were among the plethora of problematic players involved in the mystery of 2005's devastating hurricane season. One could argue that all of these elements are just pieces of Mother Nature's intricate puzzle. Yet it could be contested that global warming is causing these elements to become anomalous.

Additionally, it could be disputed that extreme seasonal or yearly meteorological phenomena are influencing the recent changes in hurricane activity. And yet again, it could be debated that these overlying phenomena are possibly being influenced by climate change, which in turn could influence hurricane activity. No matter which way one turns, science continues to elude as it gathers empirical data, and theories are derived to tie all the loose ends of countless correlations together. Nevertheless, hurricanes are really just a symptom of larger systems that could be affected by a warming world and thus cause the elements that went awry in 2005 to become more pronounced. The remainder of this chapter will address these issues.

EL NIÑO AND HIS EVIL SISTER

One needs to be careful about blaming something on global warming. Your boss for instance would likely be skeptical that global warming is causing you to sleep later, and subsequently miss work. Theoretically speaking, you could blame climate change for enhancing your slumber; perhaps higher than normal global temperatures are making you drowsier than usual. Your boss though would likely think differently on the matter; hence, your theory would have supporters (you and perhaps your coworkers) and yet your theory would have its skeptics (your boss). Still, almost anything could be blamed on global warming, and since it's a science in the making, it's hard to dispute new and emerging theories and their correlations, especially if that certain something that's to take the rap for global warming is something we don't see all too often.

El Niño is one of those rare meteorological events that, when it does occur, hugs the media spotlight as a possible sign of gloom and doom—a harbinger foretelling all that climate change may be upon us, and could become worse. Such fear mongering by the media occurred in the winter of 1997-1998 when news agencies, once learning about this weather phenomenon, spent countless hours of airtime vying for the attention of viewers around the world. El Niño, as a weather condition, was foreign to most ears before 1997, but the media changed all that. Many, who felt the media was over exaggerating a dreadful shift in weather, thought this phenomenon should be called El Hypo. After all, faith in the weathermen on our local news has not always been that high, even for a five day forecast, let alone a prediction calling for an entire season to be deluged by merciless, rapacious storms. But no one was laughing when storms of never-before-seen strength slammed the west coast of the U.S., bringing flooding, pelting rains, storm surge, and pounding surf. And what caused all of this? It was merely a couple degrees of change in the waters in the equatorial Pacific. El Niño, driven by butterfly-effect-like changes in sea surface temperatures is truly one of the most potent factors in Mother Nature's Goldilocks Complex. Yet is El Niño a sign of our temperature turning times?

El Niño has been around for a long time, yet only in recent years have we come to understand just what this weather phenomenon really is. Back in 1891, Dr. Luis Carranza, President of the Lima Geographical Society in Peru, wrote a small article about a strange countercurrent that flowed around the port of Paita, Peru. The Paita sailors, who would see this current disrupt fishing around Christmastime, coined this bizarre current El Niño, which in Spanish means "the boy child," and is also used to refer to the baby Jesus; hence, its seasonal relation to Christmas. The name stuck, and El Niño now refers to a fluctuation in water temperatures in the equatorial Pacific that occurs every few years. The opposite effect was given the misnomer of La Niña, which when translated to English means "the girl child," although this label was affixed to the antithesis of an El Niño by U.S. scientists, and not the people of Peru from which the El Niño name originated. Overtime, El Niño and its evil twin sister La Niña got a more official scientific name: the El Niño Southern Oscillation, or just ENSO for short, which refers to both El Niño and La Niña events.

The ENSO is a fascinating example of the tight coupling between our world's natural elements, and serves as an excellent illustration of how small changes in one aspect will inevitably create transformations elsewhere on our planet. The ENSO can drive the world's weather patterns and will determine just how dry or wet a region may be throughout an entire season. This is accomplished by seemingly minor fluctuations between the ocean and our atmosphere, which in turn rearranges jet streams and storms like pieces on a chessboard.

The ENSO occurs across the Equator from winds that continually blow from east to west known as the *trade winds*. These same trade winds are responsible for helping ships throughout history fill their sails with wind, pushing them quickly along the Equator from east to west. If it not for these trade winds, the Spanish would have had a much more difficult time sailing from their colonies in South America to the Philippines in the 16th century. The trade winds are the crux of the ENSO, and without them, our world would be a much different place. What influences these winds however, is a tiny trigger of Mother Nature's Goldilocks Complex that just so happens to be linked to global warming: water temperatures.

Warm water along the equatorial Pacific is blown by the trade winds from the east around Peru to the west toward Indonesia. Under normal circumstances, this leaves a large pool of warm surface water in the western part of the Pacific near Indonesia, and cooler water in the east near Peru. So much warm water gets piled up in the west that the sea level around Indonesia is actually about 1/2 meter higher than around Peru.[102] The normally warmer waters around Indonesia cause low-pressure systems to form nearby, since the warm air directly above the warm water rises. As this warm air rises in the west, cool air flows in from the east to replace it, thus creating the trade winds that blow from east to west. This provides a perplexing chicken-and-egg effect: easterly trade winds blow warm water to the west, yet the trade winds are created by the warm waters in the west. If it not for one, the other would not exist.

Under normal conditions, the pool of warm water will sit along the western portion of the Pacific, and easterly trade winds will continue to blow. If the trade winds lighten up, even just temporarily, then the whole system will go into shock. If trade winds relax a bit, then the pool of warm water starts to slosh back toward the east. As it does, it sets off a chain reaction where the trade winds, which are generated by heat rising off that pool of warm water, will continue to diminish since there is less warm water in the west to create the uprising of air for the trade winds to develop. This allows that pool of warm water to continue its abnormal easterly journey across the equatorial Pacific. Eventually, this pool of warm water will end up around Peru. When this happens, we have an El Niño.

The sea surface temperatures in that pool of warm water—like all temperature readings in a warming world—are staggeringly slight. During the intense El Niño of 1997-1998, the maximum deviation in temperature measured around the Eastern Pacific was only 2.5°C (about 4.5°F)[103] higher than normal when El Niño peaked around December of 1997. Most of the El Niños we've experienced have barely gone much above just 1°C. As insignificant as this may seem, it actually is quite dramatic. In comparison, a one-megaton H-bomb would release enough energy to heat about 1 cubic kilometer of seawater to just 1°C.[104] While that alone shows what a difference a degree can make, that's nothing compared to what it would take to heat the upper water of the equatorial Pacific by 1°C as El Niño does. This large expanse of water comprises a mind-boggling 6 *million* cubic kilometers of seawater, thus taking millions of H-bombs to heat it just one tiny tick on the ole mercury.

A La Niña is just the opposite of an El Niño where the trade winds become stronger than normal and blow too much warm water to the west, leaving the east with cooler than normal waters. In either case, these ENSO events disturb weather patterns worldwide. The Pacific is an immense body of water, and what happens there—in this world of tightly coupled systems—affects everyone everywhere in one form or another.

ENSO events displace the normal formation of high and low pressure systems, kind of like moving around large boulders in a stream of flowing water. When you move a rock in a stream, the water is diverted in different directions depending on where you place the rock. When the ENSO moves the low and high-pressure weather systems, it has a similar effect on the world's jet streams, which guide along most of Earth's storms. So when an ENSO event is underway, jet streams warp, change position, and subsequently send storms to abnormal locations. A prime example of this is how during an El Niño, the jet stream in the northern hemisphere dips southward, and sends wintertime storms that would normally hit the Pacific Northwest directly at Southern California. Although Southern California sees rain in the winter, during an El Niño year, it will see much more rain than normal as more storms are driven to the region from a lower than normal jet stream. During a La Niña, just the opposite occurs.

Other areas of our world are affected by El Niño as well. Heavy rains occur not just in Southern California, but also in Cuba, Northern Peru, Southern Brazil,

Northern Argentina, Eastern Paraguay, Bolivia, and Western Europe. On the flip side, areas that would normally see rain are affected by droughts such as Indonesia, the Philippines, Southern India, Africa and Australia. These droughts in turn often lead to widespread brushfires, thus causing further problems for these arid regions. Yet the ENSO has an interesting effect on hurricane formation, making it another intricate element in the overlying Goldilocks Complex, especially in a warming world.

During an El Niño year, hurricanes are typically not as active in the Atlantic, but they are in the Pacific. In an El Niño year, the ocean waters in the Eastern Pacific are abnormally warm, giving hurricanes in this region more fuel than they're accustomed to. During the El Niño of 1997 for instance, there were a total of 19 named tropical storms in the Eastern Pacific compared to only nine that year in the Atlantic. It's not often though that you'll hear about an active Eastern Pacific hurricane season, as most of these storms never come close to land, and are not thought of as newsworthy by the mainstream media. Nevertheless, 1997's El Niño provided enough high-octane hurricane fuel to spin up the strongest hurricane ever recorded in the Eastern Pacific: Hurricane Linda.

Linda had an impressive low of 902 millibars with winds that reached 184 mph at its peak. Linda kicked up some awesome surf for the California coast, and was responsible for sweeping some fishermen in Newport Beach off of a jetty and into the sea—all men survived though, and there were no casualties from Linda. A couple weeks later however, Hurricane Nora went racing up the coast of Baja, made landfall, and was responsible for at least two deaths and forced hundreds of people from their homes from rising floodwaters.[105]

While El Niño keeps the Eastern Pacific hurricane season quite busy with warmer than normal waters, it has a unique effect on Atlantic hurricane formation. Since El Niño warps the jet stream in the northern hemisphere southward, it places this jet low enough to blow across "hurricane alley" by the Bermuda High, and in the location where so many of the Atlantic hurricanes form. This jet stream has some very strong upper atmosphere winds that blow hard enough to increase the vertical wind shear, which as mentioned previously was not as prominent in 2005. During an El Niño year though, the upper atmosphere winds blow stronger and vertical wind shear is quite high around hurricane alley. With the strong wind shear during an El Niño, the tops of rising hurricane columns are literally blown off, thus not allowing hurricanes to form so easily. But if we enter a La Niña year, where everything is practically opposite of an El Niño season, this becomes a problem for the Atlantic coastal region.

In 2005, we were slowly drifting out of a minor El Niño, and quickly went into a neutral state right before the beginning of summer. By the end of summer, the world was caught in the grips of a La Niña, which by the end of summer 2005 had already started to raise the jet stream to a high enough latitude so as to reduce the flow of upper atmospheric winds across hurricane alley, and thus reducing the wind shear. As mentioned in the previous section, the lack of wind shear was one of the major components that allowed the 2005 Atlantic hurricanes to gain momentum and become some of the most cantankerous storms ever to pass through that region. What's puzzling about this though, is that before falling into the hands of a La Niña, only months before, the equatorial Pacific was in a state of El Niño. This rapid shift was unexpected in the early part of 2005, since ENSO events typically last anywhere from two to seven years. In the case of 2004-2005, the ENSO shift was quite sudden, occurring over a short span of only a few months.

The recent shift from El Niño to La Niña in such a short period does have some scientists scratching their heads. Yet every ENSO cycle is different, and appears to be somewhat random as well. Before the 1997 El Niño that got so much attention from the media, a powerful El Niño of nearly the same magnitude occurred 15 years prior. The El Niño of 1982-1983 had anomalous sea surface temperatures of 2.3°C, which was pretty darn close to the 2.5°C of the 1997-1998 El Niño. Yet the 1982 El Niño never gained much popularity, even though it caused heavy flooding in California, extensive droughts in the arid regions, some of the fiercest dust storms ever in Australia, a death toll of 2,000 people, and $13 billion in damages.[106] Mainstream media in 1997 however was on top of any newsworthy event at the time that could fill up their 24 hours of news broadcasting, and El Niño seemed to fit the bill, inciting enough worry amongst the population of viewers to keep ratings at an all time high.

As more and more talk of global warming hit the airwaves after 1997, murmurs of a link to El Niño and climate change began to snowball. When momentum builds on a controversial topic like global warming, especially when considering a link to El Niños, both sides of the debate tend to butt heads even more. Unfortunately, this has happened in the global warming debate, and today, uncertainties exist to tie correlations to the frequency or strength of ENSO events and global warming. There has been some research done over the past decade however, that seems to point to an increased and intensified ENSO cycle that has likely been influenced by climate change.

A reliable study conducted by Australia's Commonwealth Scientific and Industrial Research Organization (CSIRO) has found through climate models simulating the future of a warming world, that if global temperatures continue to increase, ENSO cycles could indeed become more frequent as well.[107] Additionally, there have been a number of studies conducted on coral rings, which seem to show that ENSO cycles are occurring more rapidly now than in the 19th century.[108] Reading coral rings is a lot like looking at tree rings: in times of plenty, coral grow more abundantly, thus creating a thicker amount of coral that when sliced open many years later, reveals a thicker ring. In times of naught, the rings are smaller since the coral didn't have an abundant resource of nutrients to help them grow. Analysis of coral rings in the path of El Niño along the equatorial Pacific have led to discoveries from the past 150 years that show how ocean climate has changed over time. These studies all agree that El Niños were occurring less often in the 19th century, and are becoming not only more frequent today, but intense as well. These findings also share the view of the Intergovernmental Panel on Climate Change (IPCC)[109], which reports that a more frequent, persistent, and intense ENSO cycle has occurred in recent times, especially since the mid-1970s.

Similar to a Russian doll, where each doll encases yet another doll, Mother Nature's Goldilocks Complex encompasses numerous meteorological events and phenomena. Hurricanes are like the smallest doll, merely a symptom wrapped inside of a bigger seasonal doll: El Niño. Yet even the ENSO cycles are wrapped up in an even bigger, overlying influence—a factor that spans decades, and even millennia.

THE PULSE OF TIME

In the mid 1990s, while looking for answers to explain fluctuations in fish populations, Steven Hare from the University of Washington discovered something unique about the Pacific Ocean and time. What Hare and his colleagues had found was that there were El Niño-like shifts in the Northern Pacific sea surface temperatures that had persisted for decades—warm spells that lasted for ten years or more, followed by cold spells that also lasted for a decade or more. This was puzzling, since El Niño, a phenomenon quite well known at the time, occurred in periods that would last for only a few years, and at times, La Niña would also kick in, and last for a couple of years as well. But something bigger was going on; something seemed to be superimposed on top of the ENSO events that intensified both El Niño and La Niña seasons.

Hare and his colleagues discovered what is now known as the Pacific Decadal Oscillation (PDO). Like with ENSO events, the PDO shifts ocean water temperatures in the Pacific with warm-water episodes (like an El Niño) and cold-water episodes (like a La Niña). But unlike an ENSO event, the PDO was found to last for anywhere from ten to twenty years; first there would be a warm period for a decade or two, then a cold period would follow for another decade or two.

For instance, starting in the 1930s, there was an overall increasing warm period in the Eastern Pacific that lasted until the early 1940s—about one decade or so. Then a cold water spell kicked in, and lasted for a couple of decades until around the 1970s when things then started to warm back up again in the Eastern Pacific waters.[110]

These spans of decades are far too long for an El Niño or La Niña, and during these decadal cycles, El Niños and La Niñas came and went. Steven Hare knew that this larger overlying force, this PDO, had potential for changing the climate in the same way as the smaller timescale ENSO events. And what we've found since then is that the PDO can back up an ENSO event like a big brother, adding extra muscle behind an El Niño punch.

Many other decadal oscillations have been found in the last couple decades as well, not all concentrated in the Pacific either. Some of these long period cycles, spanning decades at a time, include the Arctic Oscillation (AO), the North Atlantic Oscillation (NAO), the Interdecadal Pacific Oscillation (IPO), and the Atlantic Multidecadal Oscillation (AMO). All of these oscillations occur over long durations, showing that Mother Nature not only has a Goldilocks Complex, but also some nasty mood swings as well. Mother Nature can be perturbed by some slight variations from an El Niño, which in itself can have devastating results. But add in the fact that if Mother Nature really gets riled, she can stir up an intense cycle lasting for a decade or more to intensify the effects of oceanic phenomena.

Although our world's oceans have many decadal oscillations, the two primary events are the PDO in the Pacific and the AMO in the Atlantic—both of which can have a hefty influence on our weather and hurricane seasons as well. When the Pacific-oriented PDO is in a warm cycle, El Niño events become stronger. During a cold PDO cycle, La Niñas become the bully on the block. We can see this first hand by taking a look at a cold PDO cycle from the 1940s to the mid to late 1970s, and then compare that to a warm PDO cycle that followed it.

When the PDO was in a cold episode from the mid 1940s to around the mid 1970s, the El Niños were mild to moderate, and La Niñas were a bit more prominent, which would be expected during a cold episode in the Eastern Pacific. Then as the PDO cycle shifted and began to warm up during the 1980s, we encountered a strong El Niño in 1982, and then another strong El Niño in 1997—once again, reflecting what would be expected during a warming period. Thus, the PDO acted as an influence to El Niños and La Niñas by accentuating each of these ENSO cycles when the PDO was in their favor: warm PDOs help El Niños, cold PDOs help La Niñas.

On the other side of the planetary coin, the Atlantic-based AMO also coincides nicely with seasonal fluctuations, in particular, Atlantic hurricane data. The

AMO's link with Atlantic hurricane activity makes its correlation more of a sound fact than just a theory. During the period between the 1930s to around 1970—when the PDO was in a cold cycle—the AMO was in a warm period, basking the Western Atlantic in warmer than normal sea surface temperatures. During this warm phase, the Atlantic had some very strong hurricanes with a total of 74 storms striking the U.S. coastline. Then when a cold AMO cycle kicked in, that lasted from around 1970 to the late 1990s, there were only about 34 hurricanes that made landfall.[111] Although the warm spell from the 1930s to the early 1970s lasted nearly twice as long as the cold cycle that followed it, the ratio between these periods show that more hurricanes made landfall during the warm cycle. More importantly however, the tropical storms that did form during the warm AMO period were more intense, stronger, and costlier in damage. But the real smoking gun that ties in the AMO as an influential force on hurricane seasons, is historic data showing that twice as many tropical storms form into actual full-scale hurricanes during a warm AMO cycle than in a cold episode.[112] No matter how many hurricanes may form in any one season, an AMO warm cycle helps newly formed tropical storms to accelerate into full-blown hurricanes more easily. This was of grave concern during the 2005 hurricane season.

Aside from a few fits and starts at the turn of the new millennium, we've been seeing a fairly warm PDO cycle for nearly two decades. This has been allowing El Niños—when they do occur—to become stronger. This pattern of the PDO though, would usually cycle differently from the Atlantic's AMO cycle. For the most part, we live in a zero-sum world where everything tends to eventually equal out—or so it would seem. As mentioned in the last section on the ENSO cycles, when an El Niño occurs, there tends to be more hurricanes in the Eastern Pacific than the Western Atlantic. During a La Niña, the opposite is true. Thus, ENSO cycles tend to exhibit the zero-sum equilibrium that we'd normally expect. When comparing the PDO cycle of the Pacific to AMO cycles of the Atlantic, they kind of flip flop each other too—when the PDO is in a warm cycle, the AMO is usually in a cold cycle and vice versa. In the past, ENSO cycles tended to provide equilibrium as have the PDO and AMO cycles—when one's active the other seems dormant; when one's warm, the other is cold. What's alarming though is that recently, both the PDO and AMO seem to be in synch—both are on the rise, as though something is forcing them to work in cahoots with each other.

To gauge the sea surface temperatures and other various factors from the past, scientists use methods similar to those used in analyzing past El Niños. Although some data exists for actual temperature measurements, rings of living things serve as excellent calendars of the past. The rings in coral, trees, and even sediment layers from the ocean floor are studied, giving scientists a diary of climate events that was written and preserved by nature. By analyzing the thickness and composition of historic layers stored in the time capsules of nature, scientists have been able to successfully put together a map of time, showing how our world's ocean climate varied over not just decades, but centuries, and even millennia. Although these measurements from the past can accurately reveal events of long ago—when El Niños, PDOs, AMOs or other anomalous temperature-related phenomena occurred—the rings of the past (and other data for that matter) can't tell us when, or how long, a decadal oscillation will take place. We can however tell *when* we're in a warm or cold oscillation, and unlike the zero-sum game Mother Nature usually plays, all readings are now telling us that the PDO and AMO are heating up.

Recent AMO and PDO readings reveal that sea surface temperatures are on the rise in both the Western Atlantic and the Northeastern Pacific. This correlates with the increased frequency of El Niños since 2000 with El Niños occurring in 2002 and again in 2004. While this mostly resulted in increased rainfall and flooding in California and the Pacific Northwest, this pales in comparison to what's happened as a result of the warming AMO trend, which assisted the amplification of the 2004 and 2005 hurricane seasons. During an AMO warm cycle, it's not just the warmer than normal sea surface temperatures that are of concern; vertical wind shear is also diminished when the AMO heats up. And as mentioned earlier when talking about the hurricanes of 2004 and 2005, the reduction of vertical wind shear is a good thing for hurricanes, but bad news for us. This decrease in vertical wind shear has allowed developing hurricanes to strengthen and grow unabated. Now that we've entered into a warm AMO cycle, and all other hurricane-intensifying ingredients are prime, all eyes in the meteorological arena have turned to upcoming hurricane seasons with trepidation. With the PDO also heating up though, and not cooling down as would be expected when the AMO gets charged, many are also hearing the ringing of that global warming alarm bell once again.

When discussing *The Florida Phenomenon* in the previous chapter, I briefly mentioned how more and more people are flooding to the sunshine state despite the fact that sea levels are rising, coral reefs are getting bleached, and that hurricanes have recently been more intense. With the AMO cycle now in an upswing, *The Florida Phenomenon* could become far more worrisome. In many respects, people living along the southern U.S. eastern seaboard and the Gulf Coast have been pretty lucky for about 30 years prior to the hurricane devastation that started in 2004 and went full steam ahead in 2005. During the period from around 1970 to 2000, hurricane activity was fairly low since the AMO was in a cold cycle. During this time—a proverbial eye of the storm when fears of land-striking hurricanes waned with creeping normalcy—the total population of Florida nearly tripled, going from 6.7 million people to 17.3 million.[113] By the year 2010, it's estimated that Florida's population will grow even more to 19.6

million. This trend not only places more people in harm's way of the increasingly strong hurricanes of late, but also puts a spike in impact costs. In 1980 for example, it's estimated that there was about $1.1 trillion of at-risk property insured in Florida. But by 2005, that figure skyrocketed to $5.5 trillion.[114] When hurricanes hit Florida or the Gulf Coast states now, there is much more property to be damaged, and many more lives that could be lost.

Since we've been entering into warm cycles of both the PDO and AMO, global warming skeptics point to the fact that what we're seeing is merely just a natural cycle of good ole Mother Earth. PDO and AMO cycles have waxed and waned for millennia—this is nothing new. But recent studies[115] conducted by oceanographer David Field during his Ph.D. work at the Scripps Institution of Oceanography and as a postdoctoral fellow at the Monterey Bay Aquarium Research Institute, point to something that is hard to explain away with the decadal oscillations being just a natural fluctuation, and not correlated at all to human-induced changes to the climate. In the same way that tree and coral rings are analyzed, Field looked at sediment cores collected from the seafloor off the Southern California coast. These cores of silt, like tree and coral rings, comprise a diary of nature showing layer upon layer of time. As time passes, sediment collects on the ocean floor, and small deceased sea life collects within each sheet of silt. Inside each of the layers that Field's research team analyzed were tiny marine organisms called foraminifera (or just forams for short). These tiny forams, protected safely in the deep seafloor mud off California's coast have helped to unravel more of history's ocean temperature mysteries.

Like the limestone-forming coccolithophores mentioned in chapter 1, which breathe in carbon dioxide, manufacture a calcium coat of armor, die and then sink to the ocean floor, the forams also live comparably short lives, and when their time is up, they too descend to the ocean's depths and become part of the seabed sediment. By probing deep into the seafloor, Field and his team were able to extract thick tubular shafts of millennia-old sediment that, within each few millimeters, a record of time was revealed.

What Field found while slicing away at these long cores of seafloor mud, some as long as nine feet in length, was that the populations of the tiny forams in the Northeastern Pacific changed significantly in response to a general warming trend that started in the early 1900s. By counting the number of forams in each yearly slice of these cores, Field discovered that as ocean temperatures increased, these forams became more abundant, which is to be expected. What was astonishing though, is that covering a span of 1,400 years in his study, Field saw something quite out of the ordinary starting from around 1970 onward. From 1970 to now, the abundance of forams in Field's seafloor sediment samples was immense, matching nothing before in the entire 1,400-year time frame that was studied.

This was unexplainable by the natural variations like decadal oscillations and ENSO events of the past, leading Field to deduce that this excessive increase in forams indeed far surpassed any threshold of natural variability, including the decadal oscillations that have occurred over the past millennium.

According to Field's research, there is definitely something awry within the recent decadal oscillations, especially since the 1970s. While this is only one of many correlations that would be required for damning proof of human-induced global warming, it does thicken the plot when considering that (1) both AMO and PDO cycles are now on an upswing to a warm episode (2) global temperatures have increased (3) ENSO cycles are trending to warm and more frequent El Niños (4) sea levels have risen and (5) sea ice and glaciers have melted at a rapid pace in recent times. All of this circumstantial evidence could be viewed as pointing towards the same epoch of occurrence with a rapid acceleration of all these elements since the 1970s. Nevertheless, correlations are funny things, especially when it comes to the decadal oscillations. These long-term trends have only been researched for about twenty years or so, and there is still much to be understood about them. Until we can completely comprehend and definitively predict the decadal oscillations of our world's oceans, scientific research on the matter will remain very much a theory. But then again, this doesn't mean the subject should be swept under the rug and not considered as being a harbinger or symptom of global warming or something else requiring our attention. Instead, the decadal oscillations should serve as tips of proverbial icebergs where the uncharted territory that lies underneath needs to be further pursued to reveal its origins, progression and future.

Mother Nature most certainly has a personality of multiple dimensions. Her Goldilocks Complex is without a doubt an intricately woven fabric of the ocean, air and land. When searching for a simple answer to what appears to be a simple question about the state of our environment and its influence with a warming world, the complexity of nature's excessively tightly coupled elements eludes discovery. And when we think we've discovered all there is to know, Mother Nature humbles us with clues that show us we certainly don't understand our world as well as we'd wish. Besides the seasonal fluctuations from the ENSO and the decadal oscillations like the PDO and AMO, we're finding that there are cycles that last multiple decades, sometimes for 50-70 years, and in some cases a thousand years or more.

There is one element of nature though that we've known about for a long time—something special about the Atlantic Ocean in particular; something that drives hurricanes; something that controls Atlantic coastal temperatures; and something that is now starting to change…again.

SALT, HEAT AND THE GREAT CONVEYOR

Florida may have gained popularity in recent times, yet its landscape has changed immensely since the first European set foot in the sunshine state in 1513. This newcomer was a Spanish conquistador by the name of Juan Ponce de Leon, who on a self-funded voyage discovered a land he called *La Florida*, a Spanish way of saying *flowery*, which aptly named this flora-rich, newly charted region of the New World. Ponce de Leon's journey to reach this new tropical peninsula was not filled entirely with smooth sailing however, as he encountered a strange ocean current that at times, no matter how much wind was in his sails, would drift his boats backwards.[116] Many years later, this same current would be found to be partly responsible for much of the world's weather, tied inextricably to hurricanes, ice ages, and of all things, that decadal oscillation that's of extreme interest today: the AMO.

The strong current that held Ponce de Leon's ships in its grasp is what we now know as the Gulf Stream, which is just one piece of a much bigger system called the *thermohaline circulation*, or just THC for short. If the ocean's waters are the lifeblood of the Earth, the THC is the carotid artery. The THC helps to regulate the Earth's temperatures on a conveyor belt-like cycle through our world's oceans that, on just one lap, takes over 1,000 years to complete. Although winds drive the currents at the surface of the ocean, and early oceanographers thought this was the only current in our seas, the THC and all of its swirling components like the Gulf Stream run deep underwater, and at times flow more swiftly than just surface winds alone.

Like so many of the elements comprising Mother Nature's Goldilocks Complex, the THC is rigidly bound to the ocean's winds, sea surface temperatures, and the decadal oscillations of the AMO. The driving force of the THC however, is pumped by differences in water temperature and density within the seas themselves, making the THC a self-sustaining circulation, yet one that can collapse, and cause catastrophic results.

Throughout the oceans on our planet, water density varies widely—not just from ocean to ocean, but within each ocean as well. Density of seawater is determined by two factors: its temperature, and how salty it is. Cold water, like anything else that drops in temperature, compresses and becomes dense. The colder the water, the tighter, heavier, and denser it becomes. This is the opposite effect of thermal expansion mentioned earlier, where waters warming from climate change expand, resulting in rising sea levels. Cold water though is much denser, and is heavier drop-for-drop than warm water is. Adding to the density and weight of cold water is its salt content. The saltier the water, the heavier and denser it becomes. So cold saltwater is much heavier than warm freshwater. This is what not only sparks the TCH into its conveyor belt-like cycle, but also keeps it moving along and regulates its speed. Nature has a unique way of making this happen.

Seawater at the surface of the ocean is constantly being cooled by the wind. As wind blows over our oceans' surface, it creates a great deal of evaporation. Just like when we sweat and a breeze cools us down by evaporating the perspiration off our skin, the surface water on the ocean also drops a few degrees when wind blows over it. This evaporative cooling by itself does make the surface water on the ocean denser by making it colder, but the salinity factor is also disturbed. As the wind blows over the surface of the water, the evaporation only removes tiny molecules of water, leaving behind the salt, making the already cooled, dense surface water even denser as the salt content—for what water remains—becomes higher. This dense surface seawater becomes heavier than the less dense water below it, and begins to sink. This is the incipiency of the THC current.

Formations of sea ice also add to this collection of dense seawater, and subsequently, the push that starts the THC current flowing. As ice forms out of the cold seas of the Arctic, only the water molecules are frozen; the salt, or brine from the water is left behind. The seawater around this ice becomes even colder, just like dropping ice cubes in a glass of water. The salt left behind mixes with the dense cold water, and a mixture of heavy, cold and salty water then sinks off of the ice in the same manner that the cold, salty water left behind from wind across the surface of the ocean does. This is known as *brine exclusion*, and in a world where ice is melting at breakneck speed in the Arctic, is worrisome.

So when winds blow on the ocean and ice forms in the Arctic, the THC starts its cycle by first dropping massive amounts of cold, salty, dense water to the bottom of the ocean—just like a waterfall, but in slow motion. And just like a waterfall, once the dense water reaches the sea floor, the topography and contour of the seafloor channels this water into a flowing current that starts its journey around much of the world, being very predominant in the Atlantic where the water is much saltier than other oceans like the Northern Pacific. The THC however spans nearly the entire globe, and its circulation has worldwide impacts.

The flowing currents of the THC act like a large kind of deep-water conveyor belt where warm water is drawn from the tropical regions of the Atlantic northward to the colder Arctic. This warm water releases heat as it travels northward, and is partly responsible for keeping the coastal regions of Western Europe a couple of degrees warmer than Eastern Canada, even though they fall on the same degree of latitude. After this warm water reaches the Arctic, it eventually cools, sinks, and is taken back to the southern oceans in deepwater currents. This cycle continues day after day, year after year, decade after decade, and as mentioned earlier, takes over 1,000 years to make one full lap around the world's oceans. But the speed of this trip does vary from time to time, and when it does, our climate changes as well.

As the THC conveyor belt speeds up, the warm tropical water from the Atlantic is drawn towards the North Atlantic more quickly. On this more rapid journey northward, the THC leaves behind warmer than normal temperatures in its wake, and around its final destination as well: the Arctic. With warmer temperatures in the Arctic, sea ice melts faster. And with warmer than normal waters in the Atlantic, hurricanes gain more fuel in the southern equatorial regions as well as higher latitudes. If on the other hand the THC slows down, then warmer water is not fed to the northern regions, which would allow ice to freeze or remain frozen, and the higher latitudes would not be heated as greatly. All in all though, the THC has been pretty stable during our recent history. But that may now be changing.

Climate records indicate a correlation between a pattern of increased Atlantic Ocean winds and the recent upswing in hurricane activity in the 2004 and 2005 seasons.[117] This same data shows that during the cold AMO cycle that occurred by 1970 and lasted into the late 1990s, winds that increase the Atlantic's evaporative cooling and in turn the flow of warm water in the THC, were underway, and the effects of those increasing winds started to take hold right before the 2004 hurricane season began. It takes many years to affect changes on a cycle like the THC, which itself takes a millennium or more to complete just one journey, thus making correlations tricky when trying to time certain events of climate change to variations in the THC and its driving forces (like increased wind activity). Still, even though there is a lag on forces induced on the THC, the THC is very tightly coupled to the AMO, and in some respects, may be its forerunner, possibly causing an AMO cycle to shift from cold to warm, as seems to be the case today. The stronger winds that initially speed up the THC also reduce the difference between these lower elevation winds and the winds in the upper atmosphere, which then results in reduced vertical wind shear. The reduced wind shear, as mentioned earlier, is another hurricane helper, allowing hurricanes to grow with greater ferocity.

The THC, AMO, and hurricane formation, dance in a well-balanced choreography of nature's ballet. The right kind of winds, sped up by other elements of Mother Nature's Goldilocks Complex, sets the THC in motion. As it gains momentum, many years later the THC currents impose decadal oscillations that follow, which in turn create some prime conditions for hurricane formation. To tie in human-induced global warming to this picture is a difficult task, especially when you need to consider the impacts that global warming would have on the THC. If things continue to heat up, and more freshwater runs into the world's oceans from melting glaciers, then the salinity and density of the surface water would become lighter. Since the THC needs dense surface water to run the conveyor, the lighter freshwater melting from glaciers and pack ice would tend to slow down the THC. This would not be good either, as this has happened in the past, and will inevitably happen again. When it does, the world will change dramatically.

The THC has stopped before with the most recent halt occurring around 12,800 years ago during a 1000-year cold snap known as the Younger Dryas period, often referred to as the *Big Freeze*. During this ice age event, the glaciers built up a lot of ice, sea levels dropped, and temperatures plummeted so drastically around the summit of Greenland that temperatures there were, on the average, 27°F colder than they are today.[118] This cold snap transformed our planet's northern territories into excessively chilly regions and changed the human way of life. During the Younger Dryas, icecaps and glaciers grew and sea levels dropped as a result. Forests and grasslands died back from the cold, and species like the Irish Elk and Woolly Mammoth were driven towards final extinction. Human societies, still mainly accustomed to hunting and gathering for their sustenance, faced enormous challenges as the fruits, cereals and animals on which they depended disappeared.[119]

On the bright side, humans did adapt—they had no choice. The conditions left from the Younger Dryas cold period forced modern humans from a mode of hunting and gathering towards a new survival strategy that enabled them to live year round, particularly during the extremely long and brutal winters of the Younger Dryas when practically no food was available. This new inventive strategy consisted of cultivating and nurturing selected foods and animals that were close to their homes—an activity known today as *farming*. Many believe that although the Younger Dryas turned our world into a cold and often foreign place to live, it helped to advance modern humans from hunters and gatherers into resourceful stewards of controlled crops and livestock, which gave us a jumpstart to where we are today.

What started the Younger Dryas however was something that is alarming, since we're seeing this happening right now: a period of warming that disturbed the THC. Before every ice age, there's a warm period—a part of Mother Nature's cold-to-warm-to-cold flip-flopping cycle. Before the Younger Dryas, things got warm enough to melt a lot of ice—too much ice in fact. So much freshwater melted off of glaciers and ice caps before the Younger Dryas that the oceans were quickly filled with this less dense freshwater, which puts the brakes on the flowing of the THC. As more and more freshwater entered the Earth's oceans, there was less salty dense water to sink downward, till eventually the THC just stopped flowing.

When the THC stops moving, no warm water is transported towards the Arctic, making things especially cold in the north. When this happened some 12,800 years ago, it left the Arctic in a virtual freezer with warm temperatures remaining in the proximity of the equator. Mother Nature though, like we humans, has a way of adapting to just about anything. Over the 1000-year period during the Younger Dryas, ice was able to reform in the Arctic and in mountain-top glaciers. The forming of ice in the Arctic brought upon brine exclusion, which in turn made Arctic waters once again quite salty. Once the waters became salty again, surface water then became dense, started to sink, and the THC began to spin its wheels once again.

What history tells us, is that the THC is not at all constant, and fluctuates enough to shut itself down. Interestingly enough though, before the THC shuts down, it seems to speed up, thus increasing the warm flow of water from the equator to the Arctic. The THC then seems to hit a peak, where a rapid spiral into a shut down then occurs.

Some studies conducted in 2005 say that at least part of the THC is already slowing down.[120] While this is fraught with controversy (which I'll discuss in chapter 9) we do know that recently, the THC has given indicators that it recently sped up, which could be a temporary anomaly or a harbinger of our changing climate. Trying to predict a cycle that takes a millennium to complete is not an exact science yet, and circulations like the THC all have their slight variations from time to time. Right now though, the THC is favorably flowing in synch with the myriad elements of Mother Nature's Goldilocks Complex, including the now warm AMO cycle, vertical wind shear reduction in hurricane alley, and warming waters in the northern regions as well. Given these correlations and interactions in the Atlantic, and the plethora of problems elsewhere that are influenced by the factors of Mother Nature's Goldilocks Complex, it's not out of line to ask: *Is it Hot in Here?*

As I mentioned earlier though, the answer to this question will be quite different depending on whom you ask.

4

Global Warming, Shmobal Warming

○ ○

As I said on the Senate floor on July 28, 2003, "much of the debate over global warming is predicated on fear, rather than science." I called the threat of catastrophic global warming the "greatest hoax ever perpetrated on the American people," a statement that, to put it mildly, was not viewed kindly by environmental extremists and their elitist organizations.
—Senator James Inhofe (Okla.)

It certainly takes two to tango. In the debate on global warming, this is more than just a simple Texas two-step though; instead, there's one heck of a hoedown going on. The implications of a warming world could touch economic and political sectors, and could potentially result in drastic changes throughout societies worldwide. Because of global warming's likelihood to change the world we've all come to love, cherish, and call home, those searching for answers, as well as others responsible for critical decisions on outcomes of research can take the issue of climate change personally. This is one issue that, no matter who you are or where you live, outcomes will touch your life and the lives of your future generations as well. It's no wonder then, that both skeptics and proponents of human-induced climate change—who are just as human as the rest of us—often fight their global warming battles with fervor in a take-no-prisoners, storm-the-castle kind of way.

Global warming though raises more than just concerns of what will happen if our planet continues to heat up. Scenarios abound on what we *could* expect, and how lives would inevitably change. But what if much of these theories, correlations, and scenarios are wrong? Common sense would dictate a sigh of relief, allowing us all to wipe the sweat from our brows, and exhale a deep comforting breath. But more would be at stake if all of our worries were for naught. Although consoled that our planet were in good health, and we could all live happily ever after, if all—or even some—of the theories, correlations and scenarios were wrong, we'd all pay the price. If global warming theory were grossly in error, then enormous amounts of research funding would be in vain, countless dollars would have been drained from economies to avoid harm that never happened, and entire populations would lose faith in their leaders who led them astray.

To make science *scientific*, any theory, finding, idea, correlation or postulation of any order needs to be critiqued—the more scrutiny the better in fact. In science, this is often achieved by what are known as peer reviews, where many experts assess and dissect someone's work, preferably from a devil's advocate point of view. Double blind studies, commonly used in the research of medicines are also recognized as reliable revealers of truth, using a method where neither researchers nor the participants of the study know what is real, and what is presented as a placebo. Although a good researcher will use peer reviews and blind study cases, an excellent person of science will warmly welcome the outcomes and critiques as newfound data to further their studies and lead to a better result. Not all researchers though find this level of critique-acceptance an easy pill to swallow, and opinion can oftentimes block the light of truth from peering through on their work.

Most in the scientific community revel in healthy debates and discussions, allowing contemplation to complement deliberations, thus blazing trails of discovery with unmatched expertise gathered through collective thought, findings, and hard work accomplished by colleagues. Scientists though are—more often than not—uniquely blessed with a passive open-mindedness that keeps opinions simmering on the back burner, and not boiling over on the front of their range of research. Others though, including politicians and environmentalists are found at times to approach the global warming debate wearing blinders, searching only for fodder supporting their beliefs and agendas while filtering out chunks of truth that go against the grain of their personal presumptions. Not everyone though, whether being a person of science or not, falls into the category of bias. Yet, while sifting through the vast array of fact, fiction, formulizations, philosophies, foundations and feelings, it can be difficult to discern what is real and what is hype, and what is science, and what is not.

In this chapter and the next, I'll be talking about some of the most hotly debated topics coming from both trenches on the climate change battlefield. In this chapter, I'll be discussing some of the more outspoken skeptics on human-induced global warming as well as some of the theories disputing manmade climate change. In the chapter that follows, I'll look at the other side of the coin, and discuss some of the more popular proponent stances on this impassioned quarrel of our climate.

DNA and Dinosaurs

I was on my way to Tokyo on a long and boring flight from Los Angeles. The year was 1993, and I was headed off to the land of the rising sun for only a few days of business, and very little pleasure. I'd be coming back to Southern California in another 72 hours, giving me just enough time to meet with some company suits, attend a high-tech trade show, and gorge on some of the world's best sushi and first-rate sake.

With about nine hours of flight time on the trip over, and another eight hours on the way back, being void of a Blackberry, iPod, or laptop for that matter—back in what now seems like the Neolithic age of technology—I had plenty of time to kill. In preparation for my arduous hours of in-flight boredom on those lengthy sojourns across the Pacific, I stopped by one of the bookstores in LAX and picked up a copy of a new best seller, *Jurassic Park*, by Michael Crichton, to keep myself entertained.

Crichton, a Harvard graduate and well renowned author, had written an intriguing book on DNA and dinosaurs that had me flipping pages with enthusiasm, stunned by what I was reading and completely engulfed in Crichton's work. I couldn't put the book down; it had me mesmerized. The book, having much more detail than the movie—which I personally thought was a lame attempt of Hollywood to turn science into an action flick—was convincingly written. At the time, the company I worked for had customers that included Cray Research, and I was quite familiar with the awesome computational power of these super computers mentioned in Crichton's book. So when Crichton wrote of using those mathematical powerhouses to sequence DNA extracted from blood-sucking mosquitoes that feasted on dinosaurs millions of years ago and were now preserved in tree sap, I, at times, wasn't sure if I was reading a novel, or if this was a prophecy of science yet to come.

But alas, more than twelve years have passed since I read *Jurassic Park*, and no one has yet to clone a dinosaur. In fact, scientists are having a tough time with just common every day sheep and dogs, with numerous failed attempts with fresh, common, every day DNA, let alone something that's millions of years old buried in old amber tree sap that needs patching with amphibian DNA to fix flaws from time's erosion on the dinosaur double helixes. Back in 1993 however, as I sat on that trip from LA to Tokyo and back, I gained an utmost respect for Michael Crichton. This was not just a novelist; Crichton obviously did his homework, and had enough background in science to pull off such a magnificent work of fiction.

Michael Crichton has not only sold over 150 million books and created the television show ER, he has also run a software company, graduated summa cum laude from Harvard University, graduated with an M.D. from Harvard Medical School and has a postdoctoral fellow at the Salk Institute of Biological Sciences. Being more than just a writer, Crichton has a passion for science—in a way, a true renaissance man. Crichton's edge on technology has no doubt assisted in his ability to write numerous novels that border on the thin line between fact and fiction like *The Andromeda Strain*, *Congo*, and recently, a book on global warming: *State of Fear*.

State of Fear has become more than just another Crichton chiller thriller, and has developed into a platform for this doctor and author to voice his opinions on climate change. Although Crichton declares up front that this is a work of fiction, he does use factual footnotes, which he points out are very much real. After 620 some pages of engrossing novella folio, with a plot hinging on a hero scientist who discovers climate change is all a fraud, Crichton concludes *State of Fear* with an "Author's Message" and an appendix where he truthfully voices his stance on global warming. Although his book is a fictional work of conspiracy theories and hidden agendas of environmentalists and those in the pro global warming camp, Crichton's closing statements are compelling.

After *State of Fear* was published, many environmentalists and global warming believers, offended by the book's plot and innuendos of ulterior agendas and *parti pris*, scoffed at Crichton's work as bunk, pointing out that not only is this book a fictional yarn, but the opinions were coming from a writer, not a scientist. Crichton is however very well educated and versed in the field of medicine—although not an Earth scientist, he does have an M.D. Additionally, Crichton does admit in the "Author's Message" at the end of his book that:

> Atmospheric carbon dioxide is increasing, and human activity is the probable cause.

He then goes on to state his opinion that we are in the midst of a *natural* warming trend that began around 1850. Yet Crichton also expresses that he does,

> "...suspect that part of the observed surface warming will ultimately be attributed to human activity.

Much of what Crichton says is not to completely discredit the fact that the world may be warming; instead, his views seem deeply rooted in the fact that research up to now has not met strict enough standards. Still, he did iterate controversial statements that have earned him respect from global warming skeptics and ridicule from others such as:

> I suspect people of 2100 will be much richer than we are, consume more energy, have a smaller global population, and enjoy more wilderness than we have today. I don't think we have to worry about them.

While this kind of comment can be construed as unfounded bias, one has to consider other indubitable points Crichton makes in his ending message, including the fact that models used to predict outcomes of global warming should be used for at least ten years to prove themselves out, noting that many models can vary by as much as 400 percent. Today, myriad models are used in the science of global warming, thus letting computers with various data inputs predict the outcomes, and in some cases, simulate what has happened in the past. There is no doubt that climate change science has escalated in recent times due to astonishing findings that, in the right context, are alarming. Still, to make science *scientific*, scrutiny, and not belief should remain the focus of research. This goes for any science.

If it not for careful years of cross checking, double checking, and putting every discovery under the microscope of scientific skepticism, we wouldn't have the advances in medicine that we take for granted today. None of the FDA drugs we have today would have passed muster if it not for critical review and testing—and for good reason: our lives depend on it. When it comes to global warming, as with medicine, our lives, and the lives of every living thing on this planet are at stake. Crichton uses this as a cornerstone for his stance on the global warming issue that, in 2005 went before the United States Senate.

On September 28, 2005, Michael Crichton addressed the U.S. Senate as a guest of Senator James Inhofe from Oklahoma. In his speech[121], Crichton emphasizes the need for independent verification of science, and the dangers of politicization of research. While many misconstrued Crichton's testimony before the Senate as an opportunity to stand tall on the anti-global-warming soapbox, Crichton's speech revolved primarily around flaws in research methodologies, and how many scientists have found difficulty in replicating results of others. This, as with much of his final parting words in *State of Fear*, draws attention to the need to improve the scientific research being conducted on global warming, much of which has government funding. To drive this home, Crichton suggested to the Senate that, "Government grants should require a 'replication package' as part of funding." By providing suggestions and not just problems, Crichton displays his ability to—in many respects—remain a man of science. The idea that it should be required to have any funded global warming research project scrutinized is sound. Without having others provide their opinions and convey collective knowledge, tiny holes in one's research could quickly erode into chasms, taking science on a blind path leading to nowhere. Nevertheless, Crichton's profession is in the production of fiction, which questions his credibility as an authority on fact in the eyes of many.

Crichton closed his Senate testimony with a powerful statement that does show his concern for global warming by saying:

> In closing, I want to state emphatically that nothing in my remarks should be taken to imply that we can ignore our environment, or that we should not take climate change seriously. On the contrary, we must dramatically improve our record on environmental management. That is why a focused effort on climate science, aimed at securing sound, independently verified answers to policy questions, is so important now.

Even though Crichton expressed his concern about climate change, and he did point out that science needs to be scrutinized to bolster its findings, comments in his speech disputing recent findings brought on attacks from senators such as Hillary Clinton, who commented:

> His views on climate change are at odds with the vast majority of climate scientists; it also appears in a work of fiction…I think that the topic of this hearing is very important, but organized in a way to muddy sound science rather than clarify it.

Democratic Senator Barbara Boxer also chimed in by saying:

> We are here to talk about sound science—a worthy and important subject. We are not here to talk about plays, novels, art or music…

Michael Crichton's speech was loved by some and hated by others. Plenty of persuasive points were made not only in his testimony, but also in his message at the end of *State of Fear*. Yet Crichton's speech didn't seem to change anyone's mind. Much of Crichton's shortcomings that day on Capitol Hill had to do with three things that fall outside of the scope of whether a man who writes novels can tell the U.S. Senate how science should be run or funded.

First, anyone who did read *State of Fear* and in particular the closing "Author's Message," would have noticed Michael Crichton's final point, where he states:

Everyone has an agenda. Except me.

While this can be taken as a slap in the face to anyone viewing scientific study from an unbiased viewpoint, it's also a statement that has the potential to further degrade the credibility of anyone who purports fair and balanced research. Crichton ironically may be correct in saying that everyone has an agenda, since he himself wrote *State of Fear* with the intention to make money; this book was not an act of charity, thus putting into question his own ulterior motives for monetary gain.

The second issue causing harm to Crichton's speech before the U.S. Senate was just plain ole bad timing. By the time Crichton and other panelists testifying that day had finished their opening statements, most of the senators, including Hillary Clinton, had left to attend another Senate hearing on the ramifications on Hurricane Katrina.[122] It's hard to tell someone to remain calm and that there's nothing to worry about when chaos comes to town. While still reeling from the devastation of Katrina—a storm of great magnitude that rang the global warming alarm bell for many—Senators swayed to the pro global warming side of the debate were not exactly in the mood to hear some highly paid novelist tell them how climate change science is flawed, and cast doubt on the issue of global warming.

The third thing that blew the wind out of Crichton's sails that day was the man who invited Crichton to speak before the U.S. Senate—a staunch skeptic who has called global warming "...the greatest hoax ever perpetrated on the American people..." This man, appointed into a lofty position of representation for *we the people* is Senator James Inhofe from Oklahoma, who ironically held the position of Chairman for the Committee on Environment and Public Works, and has some damning things to say about global warming.

THE INHOFE FACTOR

While Michael Crichton appears to be a reasonable man open to new ideas and capable of changing his views upon review of proof, the man who invited him to testify in front of the U.S. Senate has a reputation for being just the opposite. Senator James Inhofe from Oklahoma is an outspoken critic on the science of global warming and he's said some downright dumb things in his day regarding the subject. Nevertheless, by reading through his lines of stump thumping rhetoric, he also has made some very interesting points that all sides of the global warming debate should consider.

Still, James Inhofe has made some bizarre remarks, especially given that he has held the position of Chairman for the Committee on Environment and Public Works—the point man and go-to-guy for much of the government's decisions on global warming research and ramifications. Nine months before bringing Michael Crichton to testify on Capitol Hill, Inhofe, in a speech before the U.S. Senate encouraged his constituents to read Crichton's fictional *State of Fear*. In his speech[123], seemingly very much convinced that Crichton's novel provided additional evidence to his beliefs, Inhofe said:

> In addition, last month, popular author Dr. Michael Crichton, who has questioned the wisdom of those who trumpet a "scientific consensus," released a new book called *State of Fear*, which is premised on the global warming debate. I'm happy to report that Dr. Crichton's new book reached #3 on the New York Times bestseller list.
>
> I highly recommend the book to all of my colleagues. Dr. Crichton, a medical doctor and scientist, very cleverly weaves a compelling presentation of the scientific facts of climate change—with ample footnotes and documentation throughout—into a gripping plot. From what I can gather, Dr. Crichton's book is designed to bring some sanity to the global warming debate

We all know that politicians can be a little whacky at times and that often they're driven by opinions and bias—aren't we all every now and then? But to recommend a work of fantasy be read and digested as plausible evidence by the U.S. Senate is more than kooky—it defies the constitution on which we stand, bringing falsehoods into an institution of sacred, moral reverence, founded by men who with intentions for freedom did not build a country out of sensationalistic novella fodder. Still, with all due respect to Senator Inhofe, his views, which often overlap with those of Michael Crichton's, deserve ample time for consideration since—although riddled with transparent bias—they drive home valid points on the validity of a very serious situation.

For many years, Senator Inhofe has been insisting that science get it right, and more importantly, that the media not hype up half-truths into apocalyptic ballyhoo. Inhofe has shown how even though scientists and media today are warning of impending disaster from an overheated planet, in the 1970s the view was far different. Inhofe points out[124] that on April 28, 1975, Newsweek printed an article titled *The Cooling World*, in which the magazine warned,

> There are ominous signs that the Earth's weather patterns have begun to change dramatically and that these changes may portend a drastic decline in food production—with serious political implications for just about every nation on Earth.

Inhofe's apprehension on global warming is also founded by other articles including one from Time magazine published on June 24, 1974 that declared:

> However widely the weather varies from place to place and time to time, when meteorologists take an average of temperatures around the globe they find that the atmosphere has been growing gradually cooler for the past three decades.

And now, a few decades later, everyone is singing a different tune, saying that the world is heating up. As Inhofe points out, "How quickly things change." The media is one outfit that without a doubt has an agenda. In the interest of monetary gain, media companies vie for your attention, thus gaining higher ratings, which in turn draw in advertising dollars, allowing the media to thrive and prosper—it's just simple economics. Yet as I expressed in the Introduction chapter of this book, the media, by grabbing hold of what's *in vogue*, proliferates a profusion of confusion. Senator James Inhofe's Doubting-Thomas stance is bolstered by this type of inaccuracy in reporting.

While the media can take a bad rap for biased portrayal of the facts, true science should not. Sadly though, science is often viewed through a window polished by the media, a portal that funnels the truth up from the realms of research to those in search of answers. The light that is finally portrayed by this prism can be distorted. Senator Inhofe's skepticism in many ways reflects a scientific paradigm in that he feels science needs to give the right answers, and not keep changing their minds. Senator Inhofe said it best in his July 2003 speech before the Senate when he stated,

> I believe it is extremely important for the future of this country that the facts and the science get a fair hearing.

Still, Inhofe's statements were spawned by media headlines and not from published scientific works. And of course, using a novelist as a witness doesn't exactly give science a fair shake by any means either.

Senator Inhofe does raise some good questions though, including one involving Earth's thermostat. Inhofe poses the query that since temperatures were likely lower at the dawn of civilization, and since we've gone through ice ages and warm periods before, what temperature would be considered ideal? Should we increase temperatures by a couple of degrees? Or should we lower them? At what temperature will Earth be comfortable? Ice ages and warm spells have ended without our help before, so why intervene now? Yet Senator Inhofe, to protect any answer that may dispute his beliefs and actually prove that increased carbon dioxide is warming our planet, in his July 2003 Senate speech said,

> What gets obscured in the global warming debate is the fact that carbon dioxide is not a pollutant. It is necessary for life. Numerous studies have shown that global warming can actually be beneficial to mankind.

> Most plants, especially wheat and rice, grow considerably better when there is more CO_2 in the atmosphere. CO_2 works like a fertilizer and higher temperatures usually further enhance the CO_2 fertilizer effect.

While it would be difficult to convince the sunburned Inuits of the north, the victims of the 2004 and 2005 hurricane seasons, and the displaced residents on disappearing islands in the South Pacific that global warming is beneficial to humankind, Senator Inhofe does speak some truth in his statements. Nevertheless, some of his eccentricity shows through, seemingly arming his arguments with back up plans of attack—just in case I'm wrong, so what?

Within mere words of *some* viable questions in his July 2003 Senate speech where he might have gained momentum with well-founded arguments, James Inhofe turns around and spouts off offensive remarks against the Intergovernmental Panel on Climate Change (IPCC). The IPCC is a leading authority on climate change research established by the World Meteorological Organization (WMO) and the United Nations Environment Programme (UNEP) involving several hundred scientists from around the world. In the next chapter, I'll discuss the IPCC in more detail. This organization does however comprise the world's largest working groups on global warming research and analysis, but this didn't stop Senator Inhofe though from saying that parts of the IPCC process,

...resembled a Soviet-style trial, in which the facts are predeter-
mined and ideological purity trumps technical and scientific
rigor.

More often than not, it's hard to win over friends and influence people by calling the workings of organizations like the IPCC, involving hundreds of scientists
worldwide with WMO and UNEP funding, "Soviet-style." Not many in the free
world wish to be referred to as communists. Surely, many scientists in the IPCC,
as well as politicians that support the IPCC were offended by Senator Inhofe's
remarks. This was further punctuated when in a July 29, 2003 hearing before the
U.S. Senate[125], Inhofe, referring to a study[126] refuting recent findings by the IPCC
that our world is abnormally warming said:

The powerful new findings of this most comprehensive study
shiver the timbers of the adrift Chicken Little crowd.

While saying this, one of the most renowned authors of IPCC research,
Michael Mann, was sitting in the same room, waiting to testify at this hearing
before the U.S. Senate. Yet Senator Inhofe didn't seem to mind mixed company,
and touted this repugnant remark anyway. Not only did Inhofe trivialize the
work of the IPCC and climate change science, but while baring his over-confi-
dence on the matter, his bias was showing. The crux of this statement though,
and a primary driver for this particular hearing, was conflicting research from the
IPCC and some skeptical scientists.

The data in question was a graph showing how the Earth has been rapidly
warming since the onset of the Industrial Revolution. This graph showed a span
of 1000 years across the bottom, with temperature readings registered by the up
and down axis displayed on the left hand side. Throughout much of history, as
this graph originally showed, the temperatures remained pretty much flat, aside
from a few minor bumps and dips here and there. This formed what looked like
a fairly flat horizontal line across much of the graph. Then near the end of the
graph, when approaching our recent time, the temperatures shot up dramatically.
Looking at the graph, a picture was revealed: a flat stick with a sharp rise. In
1998, Jerry Mahlman, a colleague of Mann who became the brunt of Inhofe's
"Chicken Little" comment before the U.S. Senate, gave this graph an unofficial
name: the "Hockey Stick Graph," aptly named by its flat horizontal line resem-
bling the stick's base, then a drastic vertical rise in temperatures imitating the
handle of a hockey stick. The name stuck, and with much of the global warming
debate being spawned from this hockey stick-like graph, it's been a point of contention ever since.

HOCKEY ON THIN ICE

Measuring global temperatures is a tough job. Imagine trying to gather temperature for one specific point in time across the entire planet. It's an arduous task to say the least. But that's only half the problem when it comes to gauging how hot our world is now compared to the past. Many years ago, people weren't walking around with thermometers taking temperature readings and recording them for later use. In fact, instrumental readings—like those taken from thermometers—have only been recorded for the past 150 years. Using this data, one could easily plot a graph showing how temperatures jostled up and down throughout these 150 years. But how does one go about gathering temperature data that goes back hundreds of years, or even thousands of years? This is a complex task, and one that stirs controversy in the global warming debate.

In previous discussions, I mentioned how rings viewed in cross-sections of coral and trees, as well as slices of seafloor sediment can reveal a diary of the past. Part of what's recorded in these natural archives is a story of how good or bad the environment was during the time a tree ring grew or layer of sediment formed. In trees and coral, when conditions for growth are favorable, the trees and coral grow better, leaving behind a thicker ring. When looking at a tree stump, or any other cross-sectional view of timber, you can count the rings—one for approximately each year—and by the thickness or thinness of each ring, you could tell what the weather was like during that year.

Tree and coral rings, sediment layers and the like are not actual temperature readings, and only act as a proxy for these readings, allowing scientists to correlate known conditions of tree and coral growth to recent temperatures. This inferred data, that's used from tree rings and such to gauge past temperatures is referred to as *proxy data*. This kind of proxy data was used in putting together Michael Mann's infamous hockey stick graph by compiling data from tree rings and other natural timepieces. Since this graph displayed temperatures for the past 1,000 years, and only the most recent 150 years could use actual instrumental data from thermometers, proxy data had to be used for the remaining 850 years of the graph. This caused concern to many skeptics including Stephen McIntyre of Toronto, Canada, a mineral-exploration consultant, and economist Ross McKitrick of the University of Guelph, Canada, who have been battling back and forth with Michael Mann for years over this hockey stick graph.

Michael Mann is a respectable man of science. He's painstakingly defended his graph, which would make a tight correlation to increased carbon dioxide emissions and a warming world—a kind of smoking gun for the pro global warming side of the debate. Mann is not new to the type of research used in constructing this hockey stick graph, and has many years of experience under his belt. He's been the organizing committee chair for the National Academy of Sciences 'Frontiers of Science' and has served as a committee member and advisor for other National Academy of Sciences panels. He served as editor for the Journal of Climate and has been a member of numerous international and U.S. scientific advisory panels and steering groups as well. Dr. Mann has been the recipient of several fellowships and prizes, including selection as one of the 50 leading visionaries in Science and Technology by Scientific American. He's received the Outstanding Scientific Publication award of the National Oceanic and Atmospheric Administration (NOAA), and recognition by the Institute for Scientific Information (ISI) for notable citation of his refereed scientific research. In August 2005, Mann was appointed Associate Professor at Pennsylvania State University in the Department of Meteorology and Earth and Environmental Systems Institute, and Director of the university's interdepartmental Earth System Science Center.[127] Despite his impressive background in science though, Mann's findings deserve critique, just like anyone else's would.

McIntyre and McKitrick originally attacked Mann's hockey stick graph by claiming that "what is almost certainly a computer programming error" in the statistical technique used by Mann and his colleagues who constructed the questionable graph, causes a single record to dominate all other records.[128] These records, taken from the proxy data, were derived from tree rings of bristlecone pines of the western United States. These pines are known to have fluctuating growth spurts, which brings in the question of the validity of their tree rings as reliable proxy data. Throw in the fact that there could be a programming error with Mann's simulations, and the whole hockey stick could be sitting on thin, cracking ice.

Mann has been debating his findings for years with McIntyre, McKitrick and others. On most of these skirmishes, Mann gains support from others who run similar simulations using other proxy data, which still reveals a hockey stick shaped graph similar to his original temperature chart made in 1998. Mann also did indeed find some programming errors in his data as McIntyre and McKitrick pointed out. Still, after making the programming corrections, the hockey stick pretty much stayed the same. Additionally, Mann found programming errors in McIntyre and McKitrick's work as well.[129]

This back and forth is what makes science, science. Without the scrutiny of peers, data would be left unproven, and possibly left with errors and numerous holes. The skepticism of McIntyre and McKitrick and others who've challenged Mann's graph has led to reinforcement of Mann's original findings over time; however, others have found some differences.

When looking over recent graphs using numerous simulations run by other groups supporting or disputing Mann's findings, the hockey stick, for the most part, remains intact. What changes is the *amount* of fluctuation in that hockey stick handle that shows how temperatures jumped by leaps and bounds since the Industrial Revolution began. What also falls into question is a period on the graph around the 15th century when Earth was transitioning from what is known as the Medieval Warm Period into the Little Ice Age—two mini-scaled fluctuations in temperature that had far reaching consequences on our planet and civilizations at the time. It's argued that there are discrepancies in Mann's graph for the 15th century temperature values[130], and that the temperatures we are seeing now are not the hottest in the past 1,000 years, and that the Medieval Warm Period was in fact hotter.

Using proxy data makes the argument a difficult one to settle. Yet whilst the bickering continues over what the exact temperature was during the 14th and 15th centuries, there is one interesting finding that most everyone overlooks: nearly all graphs, whether made by skeptics or proponents of the hockey stick, show a warming trend since the onset of the Industrial Revolution.[131] The slope of the increase—just how steep that hockey stick handle shoots up near the end of the graph—varies from graph to graph. Nevertheless, all of the findings in question show that there has been an increase in global temperatures since the onset of the Industrial Revolution as greenhouse gas concentrations became more abundant.

It's hard to make good correlations on circumstantial evidence alone. But with an overlapping consensus of data between skeptics and proponents alike, it would be safe to say that there is indeed a correlation to the onset of the Industrial Revolution and the overall rise of global temperatures. The amount of rise may be in question, but there is a rise nonetheless. It would be then safe to assert the third correlation to bind cause to effect by saying that the rise in temperatures are tied to human-induced increases of carbon dioxide. But correlations sure are funny things, especially when you consider that the Earth has gone through numerous iterations of cold and warm cycles, some that pale in comparison to what we're seeing now on any of the hockey stick graphs.

HERE WE GO AGAIN

Arguing over events that occurred during a time span of 1,000 years—as with the ongoing hockey stick graph debate—is like poring over a single grain of sand on the beach. When talking about the Industrial Revolution back in chapter 1, I made an analogy of Earth's life to a distance of one mile. With our Earth reaching the now ripe old age of 4.5 billion years (give or take a few million years here and there) and modern humans coming on to the scene only around 32,000 years ago during what's known as the Paleolithic Era, our ancient modern human ancestors showed up in just the last half inch on Earth's mile-long timescale. When looking over the last 1,000 years of Earth's history, you'd only nudge back a mere 0.03 inches on that mile-long ruler. That single grain of sand on the beach measures comparatively at about 0.03 inches, making our last 1,000 years barely a blink, and truly just a mere speck of dust on a long and winding mile-long road of Earth's evolution.[132]

Truly, our Earth has been around much longer than we have. Trying to figure out if warming that's occurred on our planet over a simple, small speck of time is a harbinger of gloom and doom does seem shortsighted. It's like asking a mayfly that lives only 24 hours the difference between summer and winter. The mayfly barely knows the difference between night and day, let alone week to week, month to month or season to season. In a way, we're like mayflies, in that we've only seen a glimpse of Earth's life. By using proxy data mentioned earlier though, we can do what mayflies can't, and look at history in detail. Scientists using proxy data have been able to analyze how Earth's climate varied not just in the last 1,000 years, but also for nearly 900,000 years or more.

During the past 900,000 years, the Earth has been shifting back and forth from extreme warm spells to tremendous cold spurts. The Younger Dryas era, mentioned earlier in chapter 3, is an example of Mother Nature's mood swings, and is just one of her many excessive temperature undulations. Given the vast array of time that these fluctuations have been occurring since the birth of Earth itself, one has to ponder the possibility that perhaps the decadal oscillations like the PDO and AMO that seem inextricably linked to our ocean currents like the THC are indeed just natural cycles of nature. If so, it could be posited that human-produced greenhouse gases may not be to blame for recent hurricane activity, the melting of polar ice, or the rise in sea levels. This however will depend greatly on what we can tell from our Earth's entire history, and not just the past 1,000 years.

The best recorded history of our planet's most recent series of ice age and warming events goes back an astonishing 130,000 years. Starting around 130,000 years ago, Earth entered a warm period known as the *Eemian interglacial*. This era lasted for about 16,000 years, with Earth experiencing similar temperatures as today's. In fact, some areas of the planet were actually warmer and moister.[133] Things had gotten so warm in fact, that sea levels during the peak of the Eemian epoch are estimated to have been at least 15 feet higher than they are now.[134] This was enough of a rise in sea level to inundate parts of Northern

Europe, leaving Scandinavia as an island. Interestingly enough though, there were far fewer humans back then, and no SUVs spewing forth greenhouse gases, even though temperatures were at least as warm as they are now.

After the Eemian epoch of warmness, a cold came over all the lands around 114,000 years ago. Following this initial cooling event, conditions often changed in sudden leaps and bounds followed by several thousand years of relatively stable climate or even a temporary reversal to warmth. Overall however, for a period of around 50,000 years, Earth became a chilly place to live. Forests in northern latitudes and high elevations retreated and grew sparse as summers became short and winters lengthened. Sea ice flourished, and glaciers formed as well. Snow that fell in high latitudes and upper elevations never had the opportunity to melt during this 50,000-year stretch and piled up from one year to the next until reaching thousands of meters thick in the coldest places on the planet.

The intense 50,000-year cold spell that followed the Eemian interglacial brought on a side effect of ice age events: dryness. Since evaporated waters were distributed mainly as snow that never melted, sea levels dropped as land was sucked dry of moisture. This escalated the destruction of forests, giving way to grasslands that require less water for growth. About 40,000 years into this cold snap (around 70,000 years ago from now), the freezing frenzy reached its peak, marking an epoch known as the *Lower Pleniglacial*. By this time, the world had become so frozen over that most of Northern Europe and Canada were covered by thick blankets of ice, hiding the once prosperous but now fallow land below them.

Then, Mother Nature woke up from her 50,000-year winter hibernation and began to warm once again, but only for a brief bit of time. For about 5,000 years, things warmed up a bit but then would drop quickly to icy conditions with wild fluctuations in temperatures that spanned centuries at a time. Then it got *really* cold again about 30,000 years ago, and temperatures continued to swing up and down for thousands of years, leading into the Younger Dryas mentioned earlier. And then once the Younger Dryas ended, we entered our current historic epoch known as the *Holocene.*

The Holocene, which we're in the midst of right now, was a welcome warmth following the Younger Dryas that, as mentioned previously, was likely a tipping point for modern humans to turn from hunter-gatherers into farmers. Once the Younger Dryas came to an end some 11,500 years ago, the Earth, in some regions, turned into a balmy place to live—quite tropical in fact. Immediately after Younger Dryas, the Earth entered several thousand years of conditions that were warmer and moister than today, known as the *Holocene Optimum.* This was an amazing time, and shocking in comparison to the ole hockey stick graph.

During the excessively warm Holocene Optimum period that occurred between 9,000 and 5,000 years ago, the Earth was a different place. Conditions were so warm and moist in some locations that the Saharan and Arabian deserts were overcome with vegetation and forests grew closer to the poles than they do now. But it is interesting to note that not all areas of our planet seemed to be affected by the warmth at the same time. In fact, while regions near the North Pole skyrocketed up nearly 4°C (about 7°F), cooling in southern latitudes prevailed. Then within a couple thousand years, the trend reversed with warmer temperatures in the south and colder temperatures in the north.[135] This is important to note, in that during the Holocene Optimum, the *average* temperatures on the planet were not overly warm, nor were they excessively cold. Today though, average temps are warm overall. Nevertheless, during the Holocene Optimum, our world was quite different than it is today.

Earth then settled from its radical pendulum swing of warmth from north to south, and cooled slightly to the temperatures we've become accustomed to in our lifetimes. From then to now, there have been two minor blips. One is called the Medieval Warm Period that occurred around the 10th century and lasted until the 14th century—a mere blink in time compared to the ice ages and warm spells of the past that lasted for tens of thousands of years. During this time, wine grapes were grown in Europe as far north as Southern Britain, and the Vikings took advantage of ice-free seas to colonize Greenland and other outlying lands of the far north. But no sooner did the Medieval Warm Period get cooking that it came to an abrupt halt.

When the Medieval Warm Period ended, a tiny epoch known as the Little Ice Age took over. This lasted from the 14th century to the mid 19th century—relatively recent actually. This 400 plus year cold snap brought brutal winters to many parts of the world, especially in Europe and North America. In the mid-17th century, glaciers in the Swiss Alps advanced, gradually engulfing farms and crushing entire villages. The River Thames and the canals and rivers of the Netherlands often frozen over during the winter, and people skated and even held frost fairs on the ice. In the winter of 1780, New York Harbor froze, allowing people to walk from Manhattan to Staten Island. Sea ice surrounding Iceland extended for miles in every direction, closing that island's harbors to shipping. The Arctic pack ice extended so far south that there are six records of Inuits landing their kayaks in Scotland! The population of Iceland fell by half as temperatures plummeted making land useless, and the Viking colonies in Greenland died out as a result.

And now, here we are, just about 150 years since the end of that last mini ice age event. Looking at Earth's wild climate of the past, before humans had coal-burning power plants and SUVs, makes one wonder what could have possibly caused such radical changes to our planet's weather system. Moreover, are we just going into another one of Mother Nature's many climatic cycles? If so, when will nature flip flop to a cold or warm spell again? Is this happening now? And how could we tell if this will happen again at some point in the future?

ONCE AROUND THE SUN

Ice ages have come and gone and warm cycles have been peppered in between. While debates rage on as to whether earthly elements such as greenhouse gases, ocean cycles, and weather phenomena are to blame for our warming world, these factors account for only 50% of the equation. We can dwell from here to eternity over hockey stick graphs and carbon dioxide readings, and bicker back and forth as to the validity of correlations made between them, but we also need to expand our horizon of thought, and look up to the stars. The remaining half of the global warming precept involves something more significant than Earth, or anything on it. The single biggest slice of the climate change pie is awarded to the heat source responsible for the warming to begin with—our Sun.

The Sun is a very hot place with a surface temperature hovering around 9,900°F, which is nothing compared to its surrounding corona which reaches temperatures near 3,000,000°F.[136] Luckily for us, we live a good distance from the Sun, somewhere in the neighborhood of 93,000,000 miles or so—far enough that it takes over eight minutes for light from the Sun to reach us. If this massive ball of nuclear fusion were located any closer or farther away, our world would be a different place. Nevertheless, without a nice healthy coating of greenhouse gases in Earth's atmosphere, life could never have taken hold here, and the heat from the Sun would merely be blown away.

The Sun, being a humongous ball of burning hydrogen and helium, does have its hiccups from time to time though, which results in differences in the energy it emits to us and other planets in our solar system. Additionally, Earth doesn't just rotate nicely in a perfectly formed circle around the Sun either, taking us at times a tad closer or farther away from this big burning orb. These two variations of the Sun's energy output, and distance from the Earth have affected the temperature of our tiny blue planet—and inevitably will again.

Sunspots are the primary culprits for the Sun's variations in energy output. These large dark spots, some measuring 50,000 miles in diameter, move across the surface of the Sun, expanding and contracting along the way. Sunspots, although dark and quite large, do not cause the Sun to shine dimmer as we might expect. Instead, the areas surrounding a sunspot are actually quite bright. So when a sunspot forms, the energy around it becomes greater and more solar energy is generated as a result. In fact, when sunspots form, massive solar flair-ups get shot out of the Sun, which can disrupt communication systems on Earth. This intense energy from sunspots though has a profound effect on the Sun's brightness. When the Sun has more sunspots, the Sun releases more energy from these spots and hence shines brighter.

Although sunspots are somewhat random and usually come and go within days at a time, the Sun does go through sunspot phases where long periods pass by with little sunspot activity, and yet other periods experience a lot of sunspot activity. This provides a sunspot cycle resulting in periods of more solar energy sent to Earth, followed by less energy radiated at our planet.

Willie Soon and Sallie Baliunas of the Harvard Observatory—two well known scientists and global warming skeptics—have compiled data showing a correlation in sunspot activity and the two recent cold-warm cycles of recent times: the Little Ice Age and the Medieval Warming Period mentioned in the previous section. Soon and Baliunas reported that when there were fewer sunspots, the Earth cooled, and that when there were more sunspots, the Earth warmed.[137] While these results seem logical, it is important to note that there has been very little data on sunspots over history, and that this topic is extremely controversial as a result. Additionally, the amount of heat variance related to sunspot activity is assumed to be relatively small compared to other factors influencing a warming world.[138] Nevertheless, the verdict is still out on sunspots, and since physics does show that sunspot activity can indeed vary our Sun's energy output, its theory and correlation to at least *some* attribution of global warming should be considered.

A much bigger effect of our Sun however is its distance from the Earth at any given time. Earth doesn't revolve around the Sun in a perfect circle, and instead follows an elliptical path. Moreover though, this elliptical path changes from time to time. Sometimes the diameter of the Earth's elliptic orbit is quite great and at other times, it's narrower. As years pass by, the Earth swings in and out of this varying ellipse, and after 100,000 years, it repeats this varying, egg-shaped pattern all over again.

A second cycle of the Earth to the Sun has to do with the way that Earth is tilted. Earth doesn't stand erect on its axis, and instead leans over a bit, titling anywhere from 21.6° to around 24.5°. To go from a tilt of 21.6° to 24.5° takes about 40,000 years. Today, the Earth's axial tilt is about 23.5°. Earth's tilt is responsible for our seasons, putting the northern hemisphere a little closer to the Sun in the summer and farther away in the winter. But when the angle of the axis tilts closer or farther from where it is now, this seasonal effect is amplified by leaning the north or south of our planet a little closer to the Sun.[139]

There is yet another significant cycle of the Sun, which occurs around every 23,000 years, known as *precession*. This cycle is a slow wobbling of the Earth as it spins on its axis. This wobbling of the Earth on its axis can be likened to a top running down, and beginning to wobble back and forth. The precession of Earth wobbles from our known direction of north pointing at Polaris (the North Star) to pointing at the fifth brightest star in the sky called Vega. In fact, since we're well into the current precession cycle, in about 10,000 years from now our North Star will be Vega, not Polaris.[140]

These three cycles of Sun to Earth are known as Milankovitch cycles, named after Milutin Milankovitch who in 1920 developed this Sun-cyclic theory based on the work of a British scientist by the name of James Croll. Croll, back in 1867, was the first to point out that major ice ages of the past might be linked with variations in the seasonal distribution of the Sun's energy. Milankovitch built on this research, and through analysis and careful examination over the years, it's been found that at least 60% of the variance in the climatic record of global ice volume on Earth comes quite close to Milankovitch's Sun-cycle predictions.[141]

While sunspots are iffy in the global warming debate, the Milankovitch cycles of the Sun are considered by most to be irrefutable evidence that our world can go through radical temperature swings not just by the perturbations of humankind or Earthly-bound elements; instead, the Sun can contribute greatly to the warmth and coolness of our planet. The Sun is indeed not constant in relation to its distance from Earth, and the amount of heat we get from that big golden orb has varied immensely over time.

Since our planet has experienced numerous Sun-Earth cycles in the past, could these now be leading us towards a warming world? And when would our next ice age hit? These questions are not so easy to answer, since, like most everything surrounding the issue of climate change, there are certainly a lot of uncertainties. From what we can tell, based primarily on the Milankovitch cycles, it looks like we're in a period of relatively small solar radiation variations with no unusual swings of the Sun over thousands of years or more. Since the Younger Dryas ended, we've been, for the most part, in what's known as an interglacial period, where things tend to warm up a bit. When working the math, taking into consideration the Milankovitch cycles and where we are now, it looks like our current interglacial period will be a bit longer than normal with the next ice age not approaching for another 50,000 years from now.

Still, we are warming. We also know we're in an extended interglacial cycle, so some warming is expected. The Sun cycles however involve steady orbits of the Earth in space, which results in a more stable increase or decrease in temperatures, and not drastic jumps or fits and starts. Providing the hockey stick graph is correct, and we've been warming anomalously fast since the Industrial

Revolution, it would be difficult to blame the even flow of orbital changes of Earth around the Sun for this dramatic climb in temperature. Nevertheless, when computing any theory, all factors and variables need to be accounted for, and can't be merely swept under the proverbial rug.

Another factor also needs consideration as well when it comes to the Sun and its relationship with our planet. We know that if it were not for the greenhouse gases trapping in the Sun's heat, Earth would be a very cold place. Water vapor is one such greenhouse gas that traps in heat from the Sun, yet clouds do a darn good job at blocking out the Sun as well, thus restricting heat from entering our atmosphere and eventually reaching the surface of Earth to warm it up. Walking outside on any cloudy day is proof that clouds can block the Sun's warmth. This is known as solar dimming, and it had profound effects following that fateful 11th day of September in 2001.

If you've ever watched a jet fly overhead, far up in the sky, you've probably seen the contrails it forms. These white clouds left behind as jets pass through the atmosphere have been thought for years by many scientists to have implications on our climate. This could never be proven however, since one would have to stop all air traffic to take measurements, and see if the lack of air traffic and the subsequent decline in contrail clouds would have any effect on temperature. The three days following the 9/11 attacks on the U.S. gave scientists this opportunity, as all flights remained grounded and no air traffic was allowed across the continental United States. During this period, scientists measured a noticeable increase in diurnal temperatures by about 1°C (almost 2°F), giving for the first time proof that contrails could indeed impact solar radiation, and subsequently climate.

Yet this blocking of the Sun, known as *solar dimming*, is not just caused by clouds and contrails from jetliners. There are far more sinister elements of nature that have caused significant dimming of the Sun, and will likely dim light from our Sun again.

Erupting Hazards

In 1878, amidst the Sunda Strait between Java and Sumatra, the Earth became restless. Series of earthquakes were jarring residents in West Java and Sumatra—a secret, foreboding call from Mother Nature that something terrible was about to occur that would change the world for a very long time. A couple years later, Northern Australia would also feel the violent shaking rise up from the depths of planet Earth, but the rumbling didn't stop there. The ground under the Krakatau edifice persisted to churn, convulse and tremble; until one day, all hell broke loose.

As mild ash and steam oozed upward from the nearby volcano of Perbuwatan in May 1883, Krakatau laid eerily silent. Then after many small eruptions, the Krakatau volcano could take no more, and on August 27, 1883, this mighty volcano made history as the world's second largest eruption ever recorded.[142] Krakatau's eruption was so intense that it was heard on Rodriguez Island some 3,000 miles away, ash fell on ships 3,700 miles away, Tsunamis reached heights of 130 feet above sea level, and 165 coastal villages were destroyed, killing over 36,000 people.[143] Krakatau's explosive eruption generated twenty times the volume of solid volcanic matter known as tephra, than was spewed forth from the famous 1980 explosive eruption of Mount St. Helens. For months after the Krakatau eruption, the world experienced brilliant sunsets and prolonged twilights due to the spread of aerosols released from Krakatau's violent eruption into the high elevations of the Earth's stratosphere. The stunning sunsets, a product of light being dazzled and distorted into hues of red and yellows from Krakatau's erupted aerosols, provided inspiration for artists, including Edvard Munch, who's famous *The Scream* displays the vibrant night sky in Norway, his native land, that was reddened by Krakatau's erupted waste. These atmosphere-bound aerosols had other profound effects on our Earth however, that went far beyond mere aesthetics. During those months of breathtaking sunsets, much of the world endured cooler temperatures as the aerosols released from Krakatau partially blocked out heat from the Sun.

Volcanoes can release massive amounts of matter into our atmosphere including greenhouse gases like carbon dioxide and sulfur-rich fumes. It's been argued by many who are skeptical of human-induced global warming that volcanoes produce more greenhouse gases than humans do. This however, is just not the case. In fact, volcanic eruptions account for only about 0.16 billion tons of carbon dioxide annually, which is diminutive compared to the 30 billion tons that humans pump into the air each year.[144] Instead of providing a heat trapping greenhouse effect, volcanoes are notorious for solar dimming, resulting in cooler temperatures from their eruptions. This dimming however will only take effect if the volcanic matter blown into the sky is thrown high enough so it's not washed away by rain in the lower layers of the atmosphere. Still, volcanoes do have an effect on climate change, but not in the way that greenhouse gases do.

Once a volcano does blow its top, tons of tephra ascend skyward. This ash can blanket the sky and block incoming light, but it's been found that the sulfur-rich gases emitted from the volcano do a better job at dimming. This was found to be true after the eruption of the Mexican volcano El Chichon in 1982. While El Chichon's eruption paled in comparison to Mount St. Helens of 1980, the Sun dimming effect created by El Chichon's eruption lowered global temperatures by three to five times as much as Mount St. Helens did. Mount St. Helens' enormous eruption resulted in a dimming that cooled the Earth by a mere 0.1°C, while El Chichon's smaller eruption dropped temperatures by as much as 0.5°C in some areas. El Chichon however, released far more sulfur-rich gas than Mount St. Helens did, which was the key to the dimming effect of these and other volcanoes. It was found that the sulfur in the eruption's gases combines with water vapor in the stratosphere to form dense clouds of tiny sulfuric acid droplets. These droplets take several years to settle out, but before they do, they decrease the temperatures in the troposphere sitting below the stratosphere, absorbing radiation from the Sun and scattering it back into space.

There have been many other volcanic eruptions in recent times that have affected global temperatures as well. The 1991 eruption of Mount Pinatubo in the Philippines produced the largest sulfur-dioxide cloud during the 20th century. This eruption, combined with an eruption one month later on Mt. Hudson in Southern Chile, diffused around the globe in a matter of months, reducing radiation from the Sun by 2%, resulting in enough dimming to cool the Earth by about 1°C for the following two years.[145]

Another impressive eruption occurred in 1783 from the Iceland volcano Laki. This eruption lasted a mind-boggling eight months, spewing forth four times more sulfuric acid than El Chichon, and 80 times more than Mount St. Helens did in 1980.[146] The climatic effects of the Laki eruption were impressive. In the eastern United States, the average winter temperature was dropped to 4.8°C below the 225-year average at the time; the estimate for the temperature decrease of the entire Northern Hemisphere is about 1°C. Benjamin Franklin, being a man who pondered the sciences, theorized that the cold conditions around 1783 were linked to the Laki eruption—and he was right.

Twenty-five years after Benjamin Franklin passed away, an eruption occurred that would have certainly astonished this celebrated statesman. In 1815, the Tambora volcano located in the archipelago of Indonesia blew its top, sending the world into a frigid cold spell that lasted well into the next year. This eruption sent a column of volcanic discharge nearly 25 miles into the sky, releasing 11 cubic miles of material high into the atmosphere along with over 200 million tons of sulfur dioxide gas that penetrated the stratosphere. The solar dimming effect from Tambora interrupted the monsoon season in India, leading to a deadly outbreak of cholera that insinuated its way across the globe. Europe experienced widespread crop failures, Ireland had its first great famine, and devastating floods hit China.[147] In North America, one year after the eruption, the ramifications of Tambora were still being felt.

History books call 1816 "the year without a summer" in North America when snow fell during June and frost was still widespread during the month of July around New England. In the middle of summer, people living in the New England states shivered, dug out their winter clothing and built roaring fires. Farmers watched helplessly as their budding fields and gardens blackened with decay. Newly shorn sheep, void of their wool coats yet sheltered in barns, perished from the cold.[148] All from the climatic effects of a volcanic eruption that occurred a year before. Yet the effects lasted only a year, and the Earth recuperated rather quickly—relatively speaking.

While volcano solar dimming does have a cooling effect on the planet, volcanic effects last typically for only a few years at a time. The sulfur-rich gases spewed forth from the world's volcanoes do not have the cumulative effect that a greenhouse gas like CO_2 does. Carbon dioxide, as you may recall from chapter 1, is recycled by various *sinks* on our planet that include plants, trees and those tiny coccolithophores in our oceans that eventually turn into limestone chalk. While nature finds it especially difficult in recent times to recycle all of the carbon dioxide emitted into the atmosphere, it has much less difficulty dealing with the fumes from volcanoes. Within a few years at most, the sulfuric byproduct from volcanic eruptions settles, and washes away. Carbon dioxide however, can remain in the atmosphere anywhere from 50 to 200 years—magnitudes longer than Sun-dimming volcanic gases.[149]

Although the Sun-dimming gases from volcanic activity dissipate within a few years, and average global temperatures are affected for usually only months or a couple years at most, recent research conducted by scientists at the Lawrence Livermore National Laboratory does show that the cooling effects on ocean water can last for many decades.[150] This study looked at ocean temperatures within years of various volcanic eruptions like Krakatau, and found that the cooling went deeper into the ocean depths than it did on land, thus leaving a much larger mass of cool water that took longer to heat up. According to the Lawrence Livermore report, this has helped to slow down rising sea levels around the globe since average temperatures are on the rise. If it not for recent volcanic eruptions like that of Mount Pinatubo in 1991 that caused solar dimming for a couple years, our sea levels would inevitably be higher than they are now as the Sun would have had a better opportunity to warm the Earth's oceans, which would then result in increased thermal expansion of the waters.

There is a worrisome note about this latest finding from the Lawrence Livermore study though. While researchers found that the eruptions of both Krakatau and Pinatubo were comparable in terms of size and intensity, and similar ocean surface cooling resulted from both eruptions, the ocean temperatures recovered much more quickly in the case of Pinatubo. This points to a troublesome cause: our oceans, being warmer around the time of the Pinatubo event were not able to stay as cool for as long as they did when Krakatau blew its top in 1883. Warmer waters in recent times provide a shorter hurdle to overcome from a Sun-dimming event to allow ocean temperatures to recover more quickly. Moreover, our oceans are heating at a faster rate than they were back in the late 1800s at the time of the Krakatau eruption, thus providing the extra kick-start needed to recover from a Sun-dimming event.

No matter what formulas can be concocted, no matter what graphs and charts can be drawn, and no matter what theories and cycles can be derived, these latest findings by the Lawrence Livermore study are yet another indicator that our planet is indeed warming up—a fact that more and more people are taking to heart.

5

Science Says

*What is a scientist after all? It is a curious man looking through a keyhole,
the keyhole of nature, trying to know what's going on.*

—*Jacques Yves Cousteau*

Without a doubt, our planet has been warming; yes, it really is getting hot in
here. Although this is not the first time our planet has gone through temperature
fluctuations, our recent increase in global temperature coincides with human
events and the physical properties of our planet that respond to increases in
greenhouse gas emissions. This isn't just my opinion though mind you; it's also
the consensus of the worldwide scientific community as a whole. As with any
controversial subject, there are indeed skeptics, yet even some of the most ardent
skeptics concede that there has been warming in recent times, and that greenhouse gases have increased as well. The arguments from the skeptics today primarily fall into one of three categories:

1. The Earth is warming but the cause is unknown.
2. The Earth is warming but mostly due to natural processes.
3. Global warming is occurring but not as much as feared.

Even Professor Richard Lindzen, an atmospheric physicist and professor of
meteorology at the prestigious Massachusetts Institute of Technology (MIT),
who is often described as the most respectable climate change skeptic, accepts the
fact that carbon dioxide and other greenhouse gases have increased due to human
activities, and that our world is warming. Yet Lindzen contends that the human-
induced impact on greenhouse gas emissions is not necessarily an important component to our changing climate. In an article Lindzen wrote for the Wall Street
Journal in 2001[151], he expressed his views by saying:

> We are quite confident (1) that global mean temperature is
> about 0.5 degrees Celsius higher than it was a century ago; (2)
> that atmospheric levels of carbon dioxide have risen over the
> past two centuries; and (3) that carbon dioxide is a greenhouse
> gas whose increase is likely to warm the earth (one of many, the
> most important being water vapor and clouds).

> But—and I cannot stress this enough—we are not in a position
> to confidently attribute past climate change to carbon dioxide or
> to forecast what the climate will be in the future.

The majority of global warming skeptics like Lindzen in the scientific community—those with a record of scholarship and that have made official statements—will bolster their skepticism on the exactness of temperature increases,
questionable data retrieval methods and the actual causes of global warming. Yet
these nonbelievers do, for the most part accept that our world is heating up and
that carbon dioxide is on the rise. Nevertheless, these skeptics only comprise a
small fraction of the world's scientific community. The worldwide scientific consensus sees things differently, believing that global warming is real and attributed
to the acts of humankind. Over the years, this consensus has grown.

The Intergovernmental Panel on Climate Change (IPCC), which is no doubt
the largest organization studying global warming today said in its Second
Assessment Report (SAR)[152] in 1995 that:

> The balance of evidence suggests that there is a discernible
> human influence on global climate.

While this was used as fodder by skeptics who'd point out that the IPCC used
the term "suggests" and didn't commit to a human-induced stance, the IPCC
strengthened its view a few years later in its 2001 Third Assessment Report
(TAR)[153] by saying:

> There is new and stronger evidence that most of the warming
> observed over the last 50 years is attributable to human activi-
> ties....In the light of new evidence and taking into account the
> remaining uncertainties, most of the observed warming over the
> last 50 years is likely to have been due to the increase in green-
> house gas concentrations.

The IPCC though is not alone in its conclusions, and in recent years; all major scientific bodies in the United States whose members' expertise bears directly on the matter have issued similar statements.[154] The National Academy of Sciences (NAS) has stated that[155]:

> Greenhouse gases are accumulating in Earth's atmosphere as a result of human activities, causing surface air temperatures and subsurface ocean temperatures to rise.

And when referring to the IPCC statements, the NAS conveys its agreement by further saying:

> The IPCC's conclusion that most of the observed warming of the last 50 years is likely to have been due to the increase in greenhouse gas concentrations accurately reflects the current thinking of the scientific community on this issue.

The American Meteorological Society (AMS) has also stated[156] their beliefs that global warming is human induced by saying:

> There is now clear evidence that the mean annual temperature at the Earth's surface, averaged over the entire globe, has been increasing in the past 200 years. There is also clear evidence that the abundance of greenhouse gases in the atmosphere has increased over the same period...Human activities have become a major source of environmental change. Of great urgency are the climate consequences of the increasing atmospheric abundance of greenhouse gases...

Many other major organizations agree that global warming is here, real, and manmade. Some of these organizations include the American Geophysical Union (AGU), the American Association for the Advancement of Science (AAAS), the U.S. National Academy of Sciences, the National Science Academies of the G8 Nations, and the U.S. National Research Council.

So many of the most well educated scientific minds in the worldwide scientific community could not be wrong—at least not entirely wrong anyway. After all, even the IPCC assessment reports do state uncertainty as to what the future holds, and this is where it all gets a bit sticky.

We know the world is warming. We also know that greenhouse gas concentrations are increasing and have climbed considerably since the onset of the

Industrial Revolution; hence, we know that a big part of the greenhouse gas increase is due to the deeds of humankind. We see the effects of a warming world with melting Artic ice and glaciers, rising sea levels, disease proliferation, and increased frequency in ENSO and other meteorological and oceanographic cycles. But that's all looking at today with a view into the past. What about tomorrow?

In the next chapter, I'll be discussing bellwethers, harbingers and other things to look for that could signal global warming progression. In the chapter following that, I'll be discussing scenarios of possible outcomes. Before we can look into the future though, we need to first look at the present, at what science is doing now, to better understand our changing climate and establish a better and clearer view of what's to come.

From Beneath the Ice

Enduring subzero temperatures at the Vostok research station in Antarctica back in the early 1970s, Russian researchers, undeterred by the icy air and gale force winds near the South Pole, persevered the brutal elements of nature and made discoveries that would change our perception of Earth's history. Tolerating bone-chilling days and nights below the seventy-degree latitudes of planet Earth, where the frigid-55°F temperatures can rob a man of his flesh should his skin be exposed to the air for mere seconds, scientists began probing the depths of the ice in search of ancient mysteries. A decade before, Greenland expeditions had found evidence under the ice and perfected methods of getting to the hidden, crystalline depths of the frozen abyss many meters below the surface. What was to be unveiled at Vostok though, would be unprecedented in the field of global climate research.

The hunt at Vostok was not for a prehistoric beast, a wooly mammoth or abominable snowman. Instead, the pursuit that day was for something no bigger than the size of a molecule or grain of sand, trapped within tiny air bubbles and frozen in time. As tiny as these treasures may be, the information they'd present would be colossal. These microscopic particles, locked tightly in layers of frozen time, would reveal astonishing details of weather throughout Earth's history, helping scientists to unravel the mysteries of global warming and shed light on dreadful predictions of Earth's future.

Unlike tree rings, coral rings, and seafloor sediment layers that disclose diaries of our past, ice does a much better job at preserving gases that can tell us not only the concentrations of things like carbon dioxide and methane throughout history, but also what the world was like at the time these gases accumulated. All one needs is a method of analyzing the content of ice, and a way of getting to the ice from eons past. Many analysis methods have been perfected over time, and luckily, our world has plenty of ice to search through.

The ice that forms over Antarctica, Greenland and various glaciers around the globe, materializes from the accumulation of snow that never melts. In places like Antarctica, where the Sun never gets much of a chance to radiate significant heat, snow that falls year after year just keeps piling up. As our world has evolved, snow has built up over Antarctica and the other frigid extremes of our planet, each year leaving behind a layer of trapped particles encapsulated in nature's own cryogenic freeze tank, held intact in a well-preserved state. As accumulating snow falls each year, it brings with it pieces of the atmosphere including oxygen, nitrogen, and a collection of greenhouse gases like carbon dioxide and methane that get trapped in tiny gas bubbles. Dust, ash, and various materials also get soaked up by the falling snow and collect in the ice as well. As years pass, and more snow falls on the region, layers of tiny gas bubbles and various debris are compressed, and locked safely away in the mounting ice. Antarctica has accumulated massive quantities of ice throughout the ages, providing scientists with a vast frozen laboratory for research.

Antarctica's ice is thick—very, very thick. This most southern continent, which is as big as the United States and Mexico combined, is covered in ice measuring thousands of meters deep. There is so much ice on Antarctica that if it were divided up, every person on Earth could have a chunk of ice larger than the Great Pyramid of Giza.[157] Antarctica, being 98% ice, holds 70% of the all the freshwater in the world, and 90% of the entire world's ice as well. Although Antarctica has an abundance of ice, it is a foreboding place to visit, survive, and conduct research, making it a monumental task to get to the buried treasures below its frigid and frosty facade.

Getting to the deep layers of ice on Antarctica and extracting the tiny gas bubbles is a tricky feat in a place where puffs of human breath can immediately crystallize and fall to the ground. Keeping an ice core hole open and protecting it from freezing over while preserving the ice core from contamination is a difficult task as well. Yet thanks to many decades of research by tireless, tenacious scientists, unremitting to the ravages of Earth's coldest regions, ice core extraction has been highly perfected over time. Scientists will first drill into the ice as they would seafloor sediment using a hollow pipe-like shaft. This allows the ice to be extracted in a long, tubular column measuring a few inches across, but many meters long. After drilling so deep (at first just a meter or two) the ice core is brought to the surface, and preserved for later inspection. The ice drill then returns to the hole, and probes to new depths, each time retrieving a meter or two of cylindrically shaped ice.

Scientists will then slice up these tubular ice cores, and based on the depth of the slice and myriad dating methods that can tell just how old the frozen water is, scientists can accurately determine the approximate year that the ice was formed. Then by melting the ice and analyzing the composition of the gases it releases, as well as the dust and ash left behind, a correlation can be made to the gases and debris, and the year they were embedded in the ice.

From analyzing ice cores, scientists can find temperature, precipitation rates, humidity, wind speed, atmospheric composition, volcanic activity and more from practically any given time in the past. The deeper the ice core, the farther back in time they can look. In the early 1970s, the Vostok ice core drilled to a depth of 950 meters, which is pretty darn impressive considering one has to drill through solid ice in a place where temperatures can reach below an astonishing-100°F, and winds can blow up to 200 mph. The ice core depth that was reached in the 1970s though increased to even greater depths over time. And recently, drilling at Antarctic sites near Vostok has exceeded 3600 meters, taking us back to 650,000 to 800,000 years ago.[158]

The ice cores have revealed a remarkable story of Earth's climate and its link to atmospheric greenhouse gases like carbon dioxide. Ice ages and warming periods are evident when charting out a timeline from the ice core data, and a trend repeated itself throughout the ages showing that when greenhouse gases like carbon dioxide increased, so did the temperatures on Earth. This data, along with data from tree rings and sediment cores, has been used to make the hockey stick graph mentioned in the previous chapter. While that graph only went back 1,000 years, the data from ice cores shows a much longer history of Earth's climate, and yet still shows the same kind of trend: as greenhouse gases increased, so did temperatures. In recent times, as shown on the hockey stick graph, greenhouse gases increased rather abruptly, and global temperatures did as well. Compared to historic events shown in ice core data dating hundreds of thousands of years ago, our recent sudden change is alarming, since in the past, drastic upswings in greenhouse gases and temperature were rare.

Another shocker that's come from the ice core data is that right now, we've beat the all-time record for carbon dioxide concentrations in the atmosphere. As mentioned back in chapter 1, before the Industrial Revolution, we had a CO_2 concentration of about 280 ppm, and today that concentration has risen to 372 ppm. Looking back over a span of 420,000 years in the Vostok ice core data, we only came close to this high level once before around 323,000 years ago when carbon dioxide concentrations got up to 300 ppm.[159] While this is alarming, bear in mind also that changes in Earth's climate history have tended to be gradual over long periods. To see our recent, abrupt increase in both greenhouse gases and temperature is somewhat disquieting, yet there were other surprising clues in the ice that took a little longer to explain, with results that were even more disheartening.

The ice retrieved and analyzed at Vostok comprises just one of many studies of ice cores conducted over the years. Ice has been sampled from sites around the world with the largest and most significant cores drilled out of sites on Greenland and Antarctica. These various ice core studies provide a consensus that jives with the hockey stick graph, showing a history of our world heating when greenhouse gases increased. What these studies also have shown however is that throughout Earth's ancient history, carbon dioxide levels would start to rise about 800 years *after* temperatures did. This almost seems counterintuitive, since the premise of today's global warming theory is that humans first increased greenhouse gas concentrations, and *then* the temperatures began to rise—and not the other way around. This 800-year lag introduces a chicken-and-egg paradox, leaving many wondering which came first—the carbon dioxide increases or the rise in temperatures. Skeptics have used this point time and again to cast doubt on the link to greenhouse gas emissions and increased global temperatures, stating—with absolute truth—that in the past, temperatures would increase *before* greenhouse gas levels rose, and not the other way around. Science though has found out why these 800-year lags occur, and the answers to this chicken-and-egg paradox tell us a couple of things about our planet, and why Earth is now warming up in an anomalous way.

The lag between rises in temperature and the increase in carbon dioxide in the past occurred during major interglacial events, primarily when major ice ages end and a warming trend then followed. To go from ice age to warmth takes about 5,000 years—it's hardly an abrupt, overnight process. The 800-year lag that is observed is only 1/6 of this 5,000-year timeframe, showing that carbon dioxide didn't cause the first 800 years of the warming, yet the remaining 4,200 years of increasing temperatures could in fact be caused by increasing levels of carbon dioxide. We also know that the cycles of Earth's orbit around the Sun—like the Milankovitch cycles mentioned previously—occur around ice age transitions, which could have sparked the initial warming of the Earth during each 800-year period before carbon dioxide levels rose. Then, as the waters in the oceans started to heat up from the natural cycle of the Sun warming our planet, carbon dioxide would be released naturally from various processes called *out-gassing*, which takes, ironically, about 800 years. It's thought that the natural releases of carbon dioxide into the atmosphere during this 800-year period then amplifies the warming process by inducing a greenhouse effect. This is known as a feedback loop, similar to holding a microphone in front of an amplifier—a cyclic loop of both cause and effect that feeds on itself and escalates the process already underway: the Earth heats naturally, 800 years later carbon dioxide is released in abundance, which causes a greenhouse effect, which then accelerates warming for another 4,200 years or so.

This explanation could indeed be used to reason why our Earth is warming now, except for two important points. First, knowing the predictability of major ice age events from Milankovitch cycles, we know that we're not due for another ice age for another 50,000 years, and that the warming period from the last ice age occurred a long time ago. So we're not entering a warming period following an ice age as is shown in the 800-year lags from the past. Secondly, the warming that we're seeing recently doesn't mimic the pattern of warming found at the ending of ice ages resulting in an 800-year lag from the carbon dioxide increases. The warming that's occurred in recent times is abrupt and closely aligns with the increases of human-produced greenhouse gas emissions, and doesn't show a natural lag. This tells us that natural forces are not at all likely to be causing the global warming we're seeing now.[160] This is a bit disturbing, in more than one way.

Although greenhouse gases like carbon dioxide and methane have varied substantially during glacial and interglacial cycles over the past 420,000 years, ice core data show that levels of these greenhouse gases were never as high as they are now.[161] From the onset of the Industrial Revolution to present day—a span of only about 250 years—there has been a 33% increase in carbon dioxide in the atmosphere, and an astonishing 150% increase in methane, bringing the concentrations of these long-lived greenhouse gases to all-time highs. Given the rapid rate that greenhouse gases have been added to the atmosphere, which has caused a quick and dramatic warming over such a short period, the historic lag in CO_2 abundance following rises in temperature now becomes more worrisome. We know that a warming world will naturally out-gas carbon dioxide from its oceans, as occurred slowly during 800-year spans before 4,200 years of greenhouse-assisted warming in the past. Yet we haven't seen the 800-year out-gassing thus far, and still our world is warming. Could our greenhouse-induced warming of recent times heat our planet enough to set off additional natural out-gassing of carbon dioxide from our oceans, thus further exacerbating our current climate change dilemma? The rapid warming of recent times, being unnatural, could set off an even greater feedback loop than what we've seen in the past, hurrying up a warming event once nature catches up and begins to out-gas more and more carbon dioxide into the atmosphere. This is a valid and frightening concern, yet one that only time can truly tell.

Antarctica is more than just an icy realm; it's also a barometer of planet Earth. We can see more than just the past from Antarctica; we can also see the present and the future as well. In this ice-covered continent, belonging to no one, not even under the sovereignty of any nation, the few residents who live there do so for scientific purposes alone. Although often stranded for months on end in bitterly cold temperatures, researchers can mingle about without fear of land predation, since polar bears remain at the other end of the planet, leaving the penguins to waddle about at will, and scientists to focus their efforts entirely on research to benefit all of humankind. This land of ice, cold, penguins and snow, provides scientists with more than just ice core samples however, and by using satellite data to measure the mass of the ice, we can also judge how well our climate is doing, and if things are warming too quickly. A recent study, using satellite imagery, provided results on the state of Antarctica, and consequently our planet as a whole. The results however were distressing.

Models simulating a warming world show that first, we'd see Arctic ice in the northern regions melt. This we've seen. The models also predict that the increased melting of northern sea ice and glaciers from Greenland and elsewhere would increase precipitation around the globe, which would eventually turn to snow and pile up on Antarctica, thus thickening the ice cap over the southernmost continent. This is something we've seen as well, up until now. Once warming increases enough, then Antarctica will also start to melt, which, according to climate models, shouldn't occur for many years. But we're now seeing this too—way too early for comfort, and something far outside the bounds of anything historically locked away in ice core data.

A study conducted from 2002-2005, and published in the journal *Science* in March 2006[162], shows that during those three years, the ice sheet mass covering Antarctica decreased significantly at a rate of 36 cubic miles per year. Earlier, from 1992-2003, the overall thickness of the ice increased, which would be expected in a warming world, and verified on climate models. The recent data showing a melting of Antarctica is alarming to say the least, since this melting was not expected for quite some time. Nevertheless, since this change is so recent, and there is so much still to learn about Antarctica, scientists are not yet dinging the global warming alarm bell just yet, but this will undoubtedly lead to more rigorous research to find out why Antarctica is starting to melt like all other icy regions on our planet.

The ice cores and other studies on Earth's frozen layers of time provide us with a good view of the past, and possibly the present. With models and simulations, we can peer into the future, although some of what we see is still a mystery; however, using data from Earth's myriad diaries as well as what we know about our world today, scientists have been able to draw some conclusions of what to expect.

THE IPCC

Hunches, inklings, gut feelings and bright ideas often provide impetus to discovery, yet an idea alone cannot be taken as fact until it's stood the test of science. Innovators and visionaries throughout history, with a small spark of thought, ignited guiding lights that have shown the way to breakthrough designs, extraordinary inventions, and formulas that formed the modern world in which we live today. Although a single mind conceived an idea, the birth of discovery and the advancement of knowledge nearly always came by way of many a helping hand. Islands of thought, isolated from the minds of others, never gain the strength and solid foundations as those that converge into continents of collaboration, combining a coalition of wit and mental might into towering pillars of intelligence and understanding.

In the scientific community, if you want to be heard, you must present your findings and data publicly for evaluation, and then continue your work based on the critique you receive. This process is known as a peer review, where one who wants to have their work taken seriously must first publish their findings and allow others to criticize, praise, or provide suggestions. Depending on the reviews one receives from his or her peers, will determine how others will later utilize this work. Research that stands tall and can prove its points will be left available for future studies by other scientists to use in their research, taking the original idea to new heights of discovery. Those who don't allow their work to be critiqued by others will be viewed as unreliable, since secrecy alludes to a lack of solid, well-proven evidence.

While scientific research provides the cornerstones to science, peer reviews supply the mortar to bind discoveries together and strengthen them over time. Discoveries found not to hold merit will eventually deteriorate, yet those that pass muster will be used as solid, steadfast foundations. Each year, thousands of viable scientific papers are written, all of them having one thing in common: a bibliography showing reference to other scientific papers that too stood the test of scientific scrutiny. Many of these papers then make their way into prestigious journals such as *Science*, *Nature*, and others, giving them a stature of credibility. These published, peer reviewed papers are then seen as being worthy of consideration within the scientific community, and have provided the field of climate change science with invaluable information as well.

The world's largest organization dedicated to the understanding of global warming does not conduct any research of its own, and instead reviews only published, peer reviewed scientific literature. This organization, which I briefly mentioned earlier, is the Intergovernmental Panel on Climate Change (IPCC). The

IPCC, led by government scientists collaborating with several hundred academic scientists and researchers around the world, has provided invaluable information on global warming since its inception in 1988, established by the World Meteorological Organization (WMO) and the United Nations Environment Programme (UNEP). The IPCC states its role as[163]:

> ...to assess on a comprehensive, objective, open and transparent basis the scientific, technical and socio-economic information relevant to understanding the scientific basis of risk of human-induced climate change, its potential impacts and options for adaptation and mitigation. IPCC reports should be neutral with respect to policy, although they may need to deal objectively with scientific, technical and socio-economic factors relevant to the application of particular policies.

The analyses conducted by the IPCC are used by policymakers around the world to determine how serious the issue of global warming is and what can be done about it. The IPCC publishes its findings every few years in assessment reports that policymakers, scientists, and just about anyone else can review and study. The IPCC's most recent Third Assessment Report, known as the TAR, was published in 2001. Like any piece of literature covering a controversial subject that is read by so many, these assessment reports have their supporters, critics and skeptics as well. While the IPCC reports hold valuable information, some of what the reports contain should not necessarily be brought up in mixed company—depending on one's views on the whole global warming debate. The IPCC's assessment reports have displayed that infamous and hotly debated hockey stick graph, which tends to rile the skeptics in the crowd. Additionally, the latest assessment report from the IPCC has some other interesting points that have raised more than just a few eyebrows as well.

In its Summary for Policymakers—an easy to understand synopsis of the TAR—the IPCC's third assessment report starts out by saying that the global average surface temperature on our planet has indeed increased over the 20th century by about 0.6°C (about 1°F). As I mentioned in earlier chapters, the difference this degree makes is astonishing, given that this is the average temperature, and that some regions have warmed even more than just a single degree. Some skeptics will argue that these temperature readings were taken near cities, which can emanate more heat than rural areas from what's known as *urban heat island effects*. Cities, where heat can be soaked up by asphalt, buildings, etc. and released like a brick oven, are indeed found to be warmer than rural areas. Nevertheless, the TAR does report that it considered this when compiling its data.

Additionally, skeptics will argue that the temperature readings were taken from only the Northern Hemisphere and not the Southern Hemisphere, which raises a valid concern.

The TAR shows two hockey-stick-like graphs of increasing temperature records: one for the past 1,000 years from only the Northern Hemisphere, and another graph showing global temperatures since 1860. Since there aren't any global instrumental temperature readings from thermometers, meters and other such devices to look through from the past millennium, proxy data like tree rings, sediment layers and ice cores, were used for the 1,000-year record beyond 1860. Data supporting this millennial graph indeed came only from the Northern Hemisphere, simply because there is abundantly more proxy data available for the Northern Hemisphere than from the Southern Hemisphere. If there were more data for the Southern Hemisphere, then one could study it. Since there isn't, no one can—reliably. Bear in mind, the IPCC only evaluates peer reviewed, published scientific literature, and has not conducted their own studies, leaving the IPCC with limited information at times that depends on who did what studies where. Besides the fact that the 1,000-year old chart reflects mostly proxy data from the Northern Hemisphere, no one has yet to show that the Southern Hemisphere averaged *colder* temperatures in the last 1,000 years, leaving this argument as only a query with no strong debatable evidence to disprove a *global* warming trend in recent times.

More importantly though is the global temperature graph in the IPCC's TAR that shows growing warmth since 1860. This graph, unlike the 1,000-year hockey stick, *was* compiled using instrumental data coming from thermometers and other such temperature-recording equipment. Moreover, this graph is a product of worldwide temperature readings—not just those taken from the Northern Hemisphere. Although this graph only goes back 140 years, like its 1,000-year-old cousin, it too shows a sharp ascent in recent temperatures, especially during the 20th century. Taking all these things and more into consideration, the IPCC consensus stands firm on its statement that global temperatures not only jumped by 0.6°C in recent times, but that the overall increase in temperature during the 20th century was *likely* to have been the largest increase of any century during the past 1,000 years. The wording of this statement though has agitated skeptics even more.

The IPCC reports are sometimes viewed as being less than a smoking gun since they are peppered with words and phrases such as *likely*, and *virtually certain*. Since the reports are premised primarily on postulations and circumstantial evidence, the IPCC protects its hide from being chewed off by skeptics and risking credibility as well by, at times, using somewhat ambiguous language. The IPCC does however point out what they mean by *likely* and other obscure terms, stating as a footnote in their assessment report that *virtually certain* means there's a greater than 99% chance that a result is true; *very likely* gives it a 90-99% chance; *likely* a 66-90% chance; *medium likelihood* a 33-66% chance; *unlikely* 10-33%; *very unlikely* 1-10%; and *exceptionally unlikely* less than a 1% chance. So when rereading the IPCC statement that says:

> New analyses of proxy data for the Northern Hemisphere indicate that the increase in temperature in the 20th century is likely to have been the largest of any century during the past 1,000 years.

...we can ascertain from the *likely* wording that some uncertainty exists in what was said. According to the IPCC statement, since it used the L-word (likely) one could feel confident that the statement holds a 66-90% chance of being fact—a view of the proverbial glass being more than half full. Nevertheless, this also means that part of that glass is indeed void of substance, leaving a 10-34% chance that the statement is wrong. Still, using the basis of the instrumental data, the IPCC does say with absolute certainty that:

> The global average surface temperature (the average of near surface air temperature over land, and sea surface temperature) has increased since 1861. Over the 20th century the increase has been 0.6 ± 0.2°C.

This statement holds no ambiguity, since they used the H-word (has). Yet some critics continue their debates by saying that this refers to only the past 140 years—a period no longer than a mere blink in Earth's climate history. In any case, through their arduous review process of data previously scrutinized by the worldwide scientific community and found credible for publication, the IPCC stands firm on its stance that the world is warming, and drives this point home as being linked to human deeds by saying:

> Concentrations of atmospheric greenhouse gases and their [warming effects] have continued to increase as a result of human activities.

As well as:

> There is new and stronger evidence that most of the warming observed over the last 50 years is attributable to human activities.

These statements, although slightly enigmatic, do stack up with other findings in the IPCC's assessment report that point to human-induced global warming correlating with butterfly-effect-like changes occurring around our world from Mother Nature's hair-trigger Goldilocks Complex. While the IPCC provides a plethora of findings on how our world is warming, it also assesses the current impacts from the changing climate. Some of these impacts include how El Niño events have been more frequent, persistent and intense since the mid-1970s compared with the previous 100 years, and that in some areas such as Africa and Asia the frequency and intensity of droughts has increased in recent decades as well.

The IPCC's third assessment report also discusses how snow cover and ice extent have decreased, stating that unquestionably, the Northern Hemisphere spring and summer extent of sea-ice has decreased by about 10 to 15% since the 1950s, and that there's been a widespread retreat of mountain glaciers in non-polar regions during the 20th century as well. Additionally, the IPCC assessment discusses how it is *likely* that there's been about a 40% decline in Arctic sea-ice thickness during late summer to early autumn in recent decades and a considerably slower decline in winter sea-ice thickness as well—something the Inuits of the north are already well aware of.

The rise in sea levels resulting from the recent temperature increases has also been confirmed in the IPCC's assessment, validating that the global average sea level has risen and ocean heat content has increased, thus inducing thermal expansion of the water in Earth's oceans. According to the IPCC's TAR, tide gauge data show that the global average sea level rose between 4 and 8 inches during the 20th century, which comes as no surprise to inhabitants of disappearing islands in the South Pacific, and the residents in and around Chesapeake Bay.

There have been other noteworthy changes in climate that are detailed in the IPCC's report as well including that it is *very likely* that precipitation has increased by 0.5 to 1% per decade in the 20th century over most mid and high latitudes of the Northern Hemisphere continents, and that it is *likely* that rainfall has increased by 0.2 to 0.3% per decade over the tropics—both of which curiously point to a warming world. Yet some things have not changed at all.

While the IPCC does report on things that have gone haywire, they do point out that a few areas of the globe have not warmed in recent decades, mainly over some parts of the Southern Hemisphere oceans and parts of Antarctica. Also, the TAR states that no significant trends of Antarctic sea-ice shrinkage or growth have been apparent since 1978; however, the IPCC's latest assessment report was issued in 2001. Since then, as mentioned earlier in this chapter, there has been a measurable reduction in Antarctic ice—something that will undoubtedly be discussed before the next assessment report is published.

While there has been some criticism regarding the IPCC's assessment reports, a strong point of contention has surfaced over the issue of hurricane activity and human-induced climate change. The IPCC, in its third assessment report clarifies that changes around the world regarding tropical and extra-tropical storms such as hurricanes, are dominated by decadal variations (e.g. the PDO and AMO), and that there have been no significant trends over the 20th century in this regard. However, the IPCC does concede that conflicting analyses has made it difficult to draw definitive conclusions about changes in storm activity, especially in the extra-tropics, thus casting a shadow of doubt on the idea of intense hurricane seasons—like that seen in 2005—being linked to global warming. This stance has caused a flurry of deliberations on both sides of the global warming debate, and as the fourth IPCC assessment report was being considered recently, things heated up beyond the boiling point in the IPCC and the scientific community as well.

THE HURRICANE CONTENTION

Chris Landsea is a brilliant man, holding the position of Research Meteorologist with the Hurricane Research Division of the Atlantic Oceanographic and Meteorological Laboratory in Miami, Florida, author of many peer-reviewed published papers, and has served as chair of the American Meteorological Society's Committee on Tropical Meteorology and Tropical Cyclones for the years 2000 to 2002. He was recipient of a National Oceanic and Atmospheric Administration (NOAA) Post-Doctoral Fellowship in Climate and Global Change for the period 1995 to 1996, and has served as an author and reviewer of the IPCC's second and third assessment reports, providing his expertise on tropical cyclones such as hurricanes and typhoons.[164] His research has been so well accepted throughout the scientific community that his work is widely cited by the IPCC. But Chris Landsea, after many years of dedication to climate change research, has left the IPCC, and in a way reminiscent of Michael Crichton's *State of Fear*. As Landsea puts it, he withdrew from participating in the next, fourth assessment report for the IPCC (known as AR4) because,

> ...I have come to view the part of the IPCC to which my expertise is relevant as having become politicized.

Crichton and his supporting Senator Inhofe were no doubt thinking, "Told you so!" But the politicization that Landsea refers to is really left up to interpretation, and has done nothing more than serve as anti-sediment fodder for the skeptics, and looked upon by many in the scientific community as overblown and melodramatic.

This whole debacle started in late 2004; shortly after that year's devastating hurricane season had ended. People were scrambling for answers and some were quick to ring the global warming alarm bell. One such bell ringer was Dr. Kevin E. Trenberth of the National Center for Atmospheric Research in Boulder—the same person who had invited Landsea to contribute to the IPCC's upcoming fourth assessment report.[165] Trenberth took part in a press conference organized by scientists at Harvard on the topic titled:

> Experts to warn global warming likely to continue spurring more outbreaks of intense hurricane activity

With all fairness to Dr. Ternberth, he did state in this press conference:

> …in the Atlantic there's no guarantee that this is going to continue, because in the Atlantic there is large, natural decade-to-decade variability in hurricane activity.…

This aligned with the IPCC's 2001 assessment report that left the link between hurricane activity and global warming open-ended. But he then added:

> …now superimposed on that natural variability is also this longer-term trend that we associate with global warming.

The global warming alarm bell rang loud and clear to the press who were listening intently for headline sensationalism. Subsequently, a flood of new stories linking global warming to the 2004 hurricane season ensued. Landsea, from hearing and reading the flurry of news coming out of this press conference viewed his colleagues like Dr. Trenberth as having "an unsupported agenda," yet what this agenda could possibly be, Landsea never really said, although he does allude to it being political.[166]

What really ruffled Landsea's feathers though was the response he received from the IPCC when complaining how a Lead Author of the IPCC could go before the public and tell the world that human-induced global warming is to blame for the powerful and damaging 2004 hurricane season. The IPCC told Landsea that Dr. Trenberth was speaking as an individual and not a representative of the IPCC, and that the media was exaggerating or misrepresenting his words.[167] This didn't satisfy Landsea, and with a mouthful of sour grapes, he left the IPCC and wrote an extensive open letter, where he concludes:

> I personally cannot in good faith continue to contribute to a process that I view as both being motivated by pre-conceived agendas and being scientifically unsound. As the IPCC leadership has seen no wrong in Dr. Trenberth's actions and have retained him as a Lead Author for the AR4, I have decided to no longer participate in the IPCC AR4.

Landsea nowadays continually objects to references made to a link between global warming and hurricane activity. And in 2005, he subjected his wrath on one report in particular that conflicts with his views, and sways toward agreement with Dr. Trenberth.

Dr. Kerry Emanuel, a professor at the esteemed Massachusetts Institute of Technology in Cambridge (MIT) wrote a paper in August of 2005 that was later published in the journal *Nature*. Emanuel's study shows a well-calculated link between increased hurricane activity and rises in sea surface temperatures.[168] By using an index that calculates storm ferocity and potential for destructiveness, Kerry shows how this index of his has notably increased since the mid-1970s, which just so happens to correlate to the same time that, according to the IPCC's TAR, El Niños increased in frequency and intensity as well. Emanuel concludes that his study correlates his calculated rising index with tropical sea surface temperature, well-documented climate signals, multi-decadal oscillations, and—here comes the real kicker—global warming. This got not only Chris Landsea steamed up; it also created a ground swell throughout the scientific community.

At long last, a prominent figure in the scientific community made a damning correlation between hurricanes and global warming. There already existed the one correlation tying global warming to rises in sea surface temperature. This has been accepted not only by the IPCC, but even by most skeptics as well. The world has warmed, so has its seas. Emanuel though inferred a cause-to-effect link, binding increased hurricane activity to increases in ocean temperature, which is caused by global warming; thus, a link was established to tie increased hurricane activity to global warming.

When Dr. Emanuel graphs his hurricane index data, it resembles that infamous hockey-stick graph, showing an upswing since the 1970s. In his published paper, Emanuel finds that the total Atlantic hurricane power dissipation—the average ferocity of hurricanes in a season—has more than doubled in the past 30 years, and in the Western Pacific, tropical cyclones have increased their power dissipation by 75%. Additionally, Emanuel reports that the duration of storms in both the North Atlantic and Western Pacific has increased by around 60%, and that the average peak wind speed of storms over the North Atlantic and Eastern and Western Pacific has also increased by about 50%.

Landsea viewed Emanuel's findings to be unfounded, and, along with Roger Pielke Jr., a fellow colleague from the University of Colorado, published rebuttals in two papers: one written for the Bulletin of the American Meteorological Society (BAM)[169] and the other in the journal *Nature*.[170] In the first article, co-written by Pielke and Landsea in BAM, an alluded attack was once again made on the *State of Fear*-like conspiracy-theory-ism towards work like Emanuel's with the following opening statement:

> Debate over climate change frequently conflates issues of science and politics.

While Pielke and Landsea's BAM article then continues on, they do concede that:

> ...since 1995 there has been an increase in the number of storms, and in particular, the number of major hurricanes in the Atlantic.

They do however argue that the changes in the past decade are not so large as to "clearly indicate" that anything other than natural cycles are at play, and that there is only "weak evidence" of a systematic increase in potential intensity.

The skeptical view of Pielke and Landsea nudged its way into Emanuel's trenches of support just a tad by agreeing that hurricanes have gotten a little stronger lately. Their view however is that it's nothing out of the ordinary. Pielke and Landsea then published separate articles for the journal *Nature*, both appearing next to each other in the same issue of this periodical. Pielke, in his article, felt that an index of actual destruction—versus Emanuel's index of wind speed, strength, etc.—should be used to confirm a theory of increased hurricane ferocity. Landsea continued his previous stance published earlier in BAM, and yet continued to concede that human-induced climate change "has the potential for slightly increasing the intensity of tropical cyclones through warming of sea surface temperatures," thus again showing a little bit of agreement on Emanuel's ideas. Immediately following both articles, a reply from Emanuel was also published.

Kerry Emanuel stood behind his original findings, and shot down Pielke's destruction index idea for the simple reason that not all hurricanes make landfall, and that over the 180-hour life of the average hurricane, their short stint over land is only around 6-12 hours—hardly enough time to judge any kind of index of intensity. Emanuel's response to Landsea was steadfast as well, as Emanuel pointed out that Landsea's view that global warming can cause a "slight" increase in hurricane strength, not only coincides with his own findings, but also is a bit under rated. Emanuel explains that, as Landsea had shown in his papers—and backed up in IPCC assessments as well—that an increase in 2°C (about 3.5°F) in sea surface temperatures in the tropics may only increase wind speeds in tropical storms by 10%. Everyone agrees on this point, yet Landsea feels this means hurricane intensity is minor; however, Emanuel says it's greater than we think. At first glance, a 10% increase in hurricane strength may not seem like much, given that sea surface temps have to increase by at least a couple of degrees. But as Emanuel pointed out, hurricanes are now lasting longer. When you plug in the numbers coming from this 10% increase in wind speed *and* the longer lifetimes of hurricanes, the increase in power dissipation climbs to 40-50%—hardly *slight* by any means of the imagination.

The debate on links over hurricane intensity to global warming is hardly settled, and it doesn't look like either side will concede 100% defeat anytime soon. Yet, that's what makes science, science—all parties giving their views to work towards a consensual outcome. It is a shame though that a brilliant man like Chris Landsea felt he should remove himself from further scientific research with the IPCC. Progress in science is made from collective collaboration, requiring that all parties respect the views of others and work together for the betterment of humankind.

The report from Kerry Emanuel wasn't the only paper on hurricane links to global warming getting attention in 2005 though. A group of Scientists from the National Center for Atmospheric Research and the Georgia Institute of Technology also published a paper in 2005 in the journal *Science* showing a direct link to recent hurricane intensification and increases in ocean water temperatures around the world.[171] Although cautiously stopping short of pinning these anomalies on global warming, in an interview with National Geographic News[172], the study's lead author, Peter Webster did say, "I'm prepared to make an attribution to global warming." While many skeptics blame the recent upsurge in hurricane activity on oscillations like the PDO and AMO, Webster's report points the finger of guilt away from multi-decadal oscillations. Since our planet tends to be a zero-sum world, where anomalous conditions in one place are seen as just the opposite someplace else, any upswing in hurricane activity say, in the Atlantic should be reciprocated by weaker activity in another part of the world. The increases found in Webster's study however, are worldwide and not just focused on a single point on the planet.

Webster's report was compiled using a 35-year analysis of the number of tropical cyclones, cyclone days, and tropical cyclone intensity. This study revealed a large increase in the number and proportion of hurricanes that reach the ultimate of strengths: categories 4 and 5. While Webster's research couldn't find any global trend in the number of actual tropical storms that have occurred over now warmer than normal ocean waters, the strength of the hurricanes has escalated. In fact, according to Webster's research, the category 4 and 5 hurricanes have almost doubled in number around the world.

The debate rages on over links to human-induced global warming and hurricane intensification, and for good reason. While melting sea ice hinders the Inuit way of life, rising sea levels are sinking small islands in the South Pacific, coral reefs die around the globe, dust blows ever more strongly from drought-affected regions, and diseases make their presence more well known in a warming world these days, hurricanes are void of creeping normalcy. Hurricanes are a violently destructive and lethal threat to not only millions of people in harm's way along the U.S. east and gulf coasts; tropical storms imperil people all around our planet. To predict a hurricane is like playing the lottery, and we only hit the jackpot if a storm stays clear of land. When it doesn't, we all lose.

THE GRAY MATTER

There is one man who for many years has challenged the stakes of betting on hurricane prediction: Dr. William Gray, Professor of Atmospheric Science at Colorado State University. For over 23 years, Gray and his team of forecasters have analyzed previous hurricane seasons and compiled predictions for upcoming periods. By looking at conditions in the past to those at present, Gray has done a fairly good job at coming up with a reasonable outlook for hurricane seasons that lie ahead. Actually, he did a pretty darn good job at forecasting what we can expect in forthcoming hurricane seasons up until 2004 and 2005, when the Atlantic fooled everyone.

Gray is one who sways to the side of doubters like Chris Landsea, Roger Pielke Jr. and Richard Lindzen, viewing a link between human-induced global warming and hurricane activity as being speculative. Gray, along with Michael Crichton, also participated on that panel back in 2005 testifying in front of the U.S. Senate on behalf of Senator Inhofe to bolster the skeptics' view on global warming. During his testimony, Gray, similar to Crichton, expressed doubts about the legitimacy of climate change science calling for verification of past studies and more rigorous scrutiny in the future. Gray passionately testified before the Senate that climate scientists "don't know much" and "have the basic physics wrong"—bold statements to make, and ones that brought on verbal attacks from Senators on the pro-global warming side of the debate. After listening to Gray's testimony, Democratic Senator Barbara Boxer from California questioned Dr. Gray's background in climate science. Gray admitted that he has not had any climate change articles published or peer reviewed—a prerequisite for credibility in the scientific community—because, as he puts it, there "is also a slight bias about accepting papers that criticize."[173] After Dr. Gray's vigorous testimony, David Sandalow from the Brookings Institution commented:

> Dr. Gray says he disagrees with [climate change science] and that he's been simmering on this topic for 20 years. I would respectfully recommend that Dr. Gray simmer his way right into the peer-reviewed scientific literature on this topic.

Gray, although being a non-peer-reviewed and non-published skeptic, is still seen by many on both sides of the global warming battlefield as somewhat credible, and not entirely crackpot-ish. Along with Landsea and Pielke, Gray has gotten a fair amount of press time in the media, and also criticized Kerry Emanuel's work, stating, "Emanuel is wrong, it's just wrong" when referring to what Gray feels are flawed measurement readings in Emanuel's data.[174] Emanuel has debated this, and stands behind his data as being accurate. Still, Gray is adamant that global warming is not changing the intensity of hurricanes, noting that between 1970 and the mid-1990s, hurricane seasons in the Atlantic were relatively quiet as levels of greenhouse gases were rising. Yet as shown on the infamous hockey stick graph, the 1970s saw cooler temperatures for a bit, and sea surface temperatures were a bit below normal as well. That all changed in recent years, evident by the upsurge on the hockey stick graph in the past decade. Nevertheless, Gray contends that the increases in hurricane activity are due to natural causes and decadal cycles like the AMO.

With 52 years worth of data, Professor Gray has many data points to use in deriving his conclusions. With a fair amount of accuracy, Gray and his team have been pretty much dead-on in hurricane forecasting.[175] By using known trends of the past, a somewhat clear picture of the future develops. For instance, when predicting the 2003 hurricane season in late 2002, Gray forecast that the Atlantic would produce twelve named storms, and sure enough, there were fourteen—pretty darn close. He also predicted that same year would see three intense hurricanes, and once again, his prediction was accurate with exactly three intense hurricanes in 2003. Prior years were also predicted with comparable accuracy.

The 2004 season was also similar with initial predictions made in late 2003 calling for 13 named storms, and sure enough 14 named storms developed in 2004. But when it came to intense storms, his initial prediction for 2004 came in at only three, when that year actually produced twice as many monster hurricanes.

And then, 2005 got the best of everyone, including Professor Gray. His prediction in late 2004 for only eleven named storms in 2005 was way off compared to the actual 27. And as for intense hurricanes, his prediction of only three was far surpassed by an occurrence of an astonishing seven massive storms—more than twice that of his original forecast. Gray does tighten up his forecasts as the clock ticks down and months get closer to the actual hurricane season, yet even in his updated report in the beginning of October of 2005, he had only predicted 20 named storms—still seven shy of the grand tally that year.

With Gray's forecasts—based primarily on historical data—being thrown into mayhem in 2004 and more so in 2005, one has to wonder just how reliable historical storm data really is. Hurricanes are complex storms. Linking them to global warming is even more complicated. Basing one's findings solely on historic data though, without accounting for the future of a warming world and the ramifications of climate change may not be the way to go anymore. Perhaps we should be looking towards new research and studies that don't turn a blind eye to human-induced climate change.

The published, peer-reviewed studies of Kerry Emanuel and the research team from the National Center for Atmospheric Research and the Georgia Institute of Technology that were presented in 2005, do show convincing theories tying increased hurricane ferocity to a warming world, and could indeed be used as variables in future predictions. Even Gray's forecasts have been up-scaled when taking into account the warmer than normal sea surface temperatures as of late. Moreover, even most skeptics believe that our world is warming, as are the oceans, which increases the fuel supply for hurricanes. In some respects, Professor Gray's studies do correlate with the pro-global warming stance and the links to upswings in hurricane intensity, since his forecasts take into account sea surface temperatures that have indeed been warmer than normal. But as with other skeptics, Gray doesn't seem to want to budge on his beliefs, or utilize findings from those that disagree with him. It would seem logical that if one has solid evidence to dispute the findings of others, then that certain someone would argue the matter in the public scientific arena, and provide his or her findings for peer review. Gray however, refuses to do so, and publishes his findings in his own personal forum. While Dr. Gray's predictions became riddled with inaccuracies for the intense 2004 and 2005 hurricane seasons, it would stand to reason that not only could his studies—and hence foundations—be flawed, but he perhaps should also work with others to find out why.

For instance, Gray asserts that multi-decadal oscillations like the AMO and a more active thermohaline circulation (the THC mentioned in chapter 3) have an influence on the recent increase in hurricane activity. Yet he feels this is not at all linked to global warming. Gray points out that the THC is driven by variations in the salinity of the ocean water. But, as I mentioned in chapter 3 when discussing the THC, the salt content of the oceans can indeed be fluctuated by increased amounts of freshwater, which can come from melting glaciers and ice. Since freshwater ice is melting around the globe from a now warmer than normal world, one would think that this should be taken into consideration when discussing any effect from the THC and its role in hurricane seasons. Moreover, global warming predictions are calling for Earth's temperatures to continue climbing, which would result in more meltwater flooding off of glaciers that could continue to affect the THC. Yet Gray has chosen not to mention the possibilities of global warming effects on the THC or anything else related to hurricane activity in his forecast reports. If his papers were peer-reviewed though, surely he would receive feedback such as this to help him tighten up his forecasts by taking into account global-warming-type impacts. But he doesn't.

Professor Gray, Michael Crichton, Chris Landsea, Roger Pielke Jr. and Richard Lindzen, have all made an exceptional case against global warming effects on our planet. Although none of them has yet to pull out a smoking gun, their contentions have assisted the scientific community in better establishing facts by being challenged to get it right. As the challenged global warming researchers have continued their quest for answers, the evidence keeps mounting ever higher that our world is warming up, and humankind is at least partially to blame. There are many consequences, side effects and global ramifications from a warming world. While it is still highly debated as to whether or not recent hurricanes are one such reaction to a heated planet, consensus still agrees that humankind, and the emissions of our greenhouse gases are changing our world.

THOSE IN AGREEMENT

Skeptics have and continue to do a great job at helping the scientific community sort fact from fiction. So much is still unknown about the world we live in, with mysteries that can only be revealed through arduous, scrutinized research. Nevertheless, the evidence is mounting up on the pro global warming side of the debate, showing that even though there are many uncertainties, there is indeed consensus in the scientific community that global warming is real, and that it stems from the increase in human-induced greenhouse gas emissions.

In the beginning of this chapter, I briefly discussed a number of renowned organizations that share similar views as the IPCC, agreeing with the human-induced global warming consensus. What I didn't mention however, is just how widespread this consensus is. The funny thing is, with such extensive agreement on the causes of global warming, it still has not been completely embraced by some, including those in high places. Skeptics and believers alike can grab the media spotlight from time to time, but only tips of scientific icebergs are ever revealed. And only when a topic aligns with the media's story *du jour* and is seen as *in vogue*, do we the people ever hear about it from the news crews to realize what is really going on. The truth is, there is far more consensus that human-induced global warming is real than there are skeptics who discount the theory wholeheartedly. The media can argue global warming issues from here to eternity, but only when it fits within the benefit of their ratings will they tell us the real *going's on*. I though, want to convey to you all that is true and that which is not—something that goes far beyond a two-minute news segment on TV, or an eye-catching sensationalistic front-page headline.

Before closing out this chapter on the scientific view of global warming, I thought it best to highlight the momentum driving the pro-global warming view. Since consensus alone is void of novelty, it's often viewed by the media as having a bland, unexciting, non-ratings-grabbing appeal. As such, the pro-global warming consensus is oftentimes not found on front-page headlines and can tend to be overlooked. But the fact remains that compared to the extremely small number of skeptics, there are far more people in the field of science, research and even politics who feel global warming is real enough that action should be taken to avert disaster. Are there hidden agendas here? Possibly. Yet it would be hard to believe that so many people from around the globe are working together on some hidden, secret scheme, especially one that is geared towards saving lives, improving the human way of life, the overall survival of humankind, and the wellbeing of the only planet in the universe that we have at our disposal.

The Group of Eight nations (G8) consists of Canada, France, Germany, Italy, Japan, the United Kingdom, the United States of America, and the Russian Federation. This large gathering of countries meets from time to time to discuss topics such as health, law enforcement, labor, and issues of mutual or global concern. The National Science Academies of the G8 comprise some of the most prominent institutions in the world including the Royal Society of Canada, the French Academié des Sciences, Germany's Deutsche Akademie der Naturforscher, Accademia dei Lincei of Italy, the Science Council of Japan, the

Russian Academy of Sciences, the UK's Royal Society, and the National Academy of Sciences of the United States. Together, these institutions, working *pro bono*, address critical national issues and give advice to federal governments and the public as well. Some of the issues the National Academies address include such things as agriculture, food and nutrition to feed our world, biological sciences and impacts to societies, engineering and technology that can impact human life, health and medicine including HIV and AIDS policies, astrophysics and planetary exploration, transportation and more. The National Academies of the G8 run the full gambit of scientific studies, and being a *pro bono* organization, they've provided stoic efforts for the betterment of humankind. In addition to the myriad studies and issues addressed by the National Academies, one topic in particular has received recent attention: global warming.

The National Academies' collaboration of brilliant minds from around the globe has stated their concern for global warming, and the wellbeing of our planet and people everywhere affected by climate change. The G8 National Academies are in agreement with the IPCC view that human-induced global warming is real, and along with members from the Chinese Academy of Sciences, India's National Science Academy and Academia Brasiliera de Ciências of Brazil, signed a statement[176] in June 2005 to this effect. In this statement, the G8 Academies of Science, along with those from Brazil, China and India, stress that the scientific understanding of climate change is now sufficiently clear to justify nations and world leaders to take prompt action. In this statement, the stance of the National Academies is evident from their opening remarks:

> There will always be uncertainty in understanding a system as complex as the world's climate. However there is now strong evidence that significant global warming is occurring. The evidence comes from direct measurements of rising surface air temperatures and subsurface ocean temperatures and from phenomena such as increases in average global sea levels, retreating glaciers, and changes to many physical and biological systems. It is likely that most of the warming in recent decades can be attributed to human activities (IPCC 2001). This warming has already led to changes in the Earth's climate.

Earlier, in 2001, an arm of the United States National Academy of Sciences (part of the G8 academies) known as the United States National Research Council, published a report[177] that explicitly endorses the IPCC's view of climate change, and stated:

> The IPCC's conclusion that most of the observed warming of
> the last 50 years is likely to have been due to the increase in
> greenhouse gas concentrations accurately reflects the current
> thinking of the scientific community on this issue…. Despite
> the uncertainties, there is general agreement that the observed
> warming is real and particularly strong within the past 20 years.

The consensus of the National Academies time and again agrees with the pro-global warming consensus. Yet, they are not the only major organization to take this stance. The American Meteorological Society (AMS), well known as setting the standards for weather reporting and meteorological study, issued an executive summary[178] in 2003 that placed their stake in the ground, marking their stance that they too agree with the IPCC. In this report, which I mentioned briefly in the beginning of this chapter, the AMS states that:

> There is now clear evidence that the mean annual temperature
> at the Earth's surface, averaged over the entire globe, has been
> increasing in the past 200 years.

This quote, favoring and supporting the infamous hockey stick graphs that have been so hotly debated by the skeptics, is not just a report from an individual from the AMS mind you; this is quoted from their policy, cast in stone in 2003. Embracing this paradigm of planetary warming due to the deeds of humankind, the AMS executive summary states their concern by saying:

> Human activities have become a major source of environmental
> change. Of great urgency are the climate consequences of the
> increasing atmospheric abundance of greenhouse gases and other
> trace constituents resulting primarily from energy use, agricul-
> ture, and land clearing. These [gases and elements] interact
> strongly with the Earth's energy balance, resulting in the prospect
> of significant global warming….For many nations, the possibil-
> ity of substantial climate change is viewed as likely to have a seri-
> ous impact on the global environment and on human welfare
> over the course of the next few decades to centuries. Because
> greenhouse gases continue to increase, we are, in effect, conduct-
> ing a global climate experiment, neither planned nor controlled,
> the results of which may present unprecedented challenges to our
> wisdom and foresight as well as have significant impacts on our
> natural and societal systems. It is a long-term problem that

requires a long-term perspective. Important decisions confront current and future national and world leaders.

The consensus reaches much farther than these large, well-known organizations though, with common sentiment felt throughout the scientific community as a whole. In December 2004, the journal *Science* published an essay[179] written by Naomi Oreskes, a PhD from Stanford, and Associate Professor in the Department of History and the Program in Science Studies at the University of California, San Diego. In this essay, Oreskes summarized a study of scientific literature on climate change where, by searching through myriad peer-reviewed papers, she could see what stance each author had taken on global warming. Oreskes analyzed 928 abstracts of papers from refereed scientific journals between 1993 and 2003, listed with the keywords "global climate change." Out of these 928 papers, written by a plethora of peer-reviewed scientists, 75% of the papers expressed explicit endorsement of the pro-global warming consensus position—an overwhelming majority. Remarkably, what was even more astonishing is that none of the papers disagreed with the consensus position that global warming is real and human-induced.

Could so many of the world's leading scientists be wrong? Could all of these people have hidden agendas? Not likely. Prejudiced plots and *parti pris* make for enticing novella fodder and are no doubt a compelling draw at the box office. But to believe that all of these scientists are in cahoots with each other is either plain ole denial, or baseless paranoia. Our government couldn't possibly hide such a vast scheme, neither could world nations and their scientists. Heck, the U.S. government can't even find the world's most popular terrorist (Osama Bin Laden), or detect if a country actually has weapons of mass destruction or not. God bless 'em, but the U.S. government is just not bright enough to pull off such a grandiose scheme that would fool the world into thinking that global warming is real for some kind of ulterior motive. In fact, you'd think it would be the opposite, where the U.S. government would want to pooh pooh the whole global warming thing, so as not to alarm the American people, which could result in disastrous economic ramifications. To think then that scientists, most of whom entered their fields of research in altruistic fashion, would be so cloak-and-dagger as to make up the whole warming-world idea and hide secrets from the American people and the entire global community for some kind of ulterior motive just seems a bit far-fetched.

There is no doubt that uncertainties loom on the horizon of climate change research. We don't have all the answers, yet we know our world is heating up. As it does, there are certain things we need to be aware of.

6

Bellwethers

Change is the law of life. And those who look only to the past or present are certain to miss the future.

—*John F. Kennedy*

Whether we like it or not, the future is headed our way. Time, as we know it in this universe, travels in one direction only—forward, and often at the speed of light. As time marches on, our world continually changes and in a warming world, changes are inevitable. Preparation is paramount when it comes to global warming, but is there really much we need to be concerned over? After doing about a year of research for this book, and digging through myriad scientific papers, studies, and facts while fishing through the white noise and ramblings of crackpots on both sides of the climate change debate, I can honestly say that I am a bit alarmed, yet I sleep well every night. After all, my ancestors of long ago lived in a much more foreboding place than we do now, and as time trampled on, so did our innate human ability to conqueror the unknown, fight and endure the most dreadful of diseases, and just plain adapt to whatever nature threw our way.

Our world, in many respects, is not the easiest place in which to live. Microscopic viruses and bacteria can fell the strongest of we humans. We're definitely not the top of the food chain, evidenced by entering ocean waters or wilderness unarmed or protected. With over 70% of the world's surface covered in water, we hardly have any dry land on which to plant our feet. Moreover, cold temperatures prevent us from inhabiting the extreme reaches of our planet, and high altitudes can do the same. The burning sands of deserts thwart our attempts of survival, and even the water from Earth's vast oceans cannot be consumed, lest we suffer the ravages of saltwater poisoning. Storms can ravage not only coastlines, but thunderous cells moving across open land can whip up fierce winds that uproot trees and houses. Bolts of static electricity can flog the ground, at times killing innocent victims and setting forests ablaze. And seasons can change within months from balmy warm to freezing cold, with extremes that pose health risks from heat exhaustion to frostbite. Our world may seem like nirvana, but it is, has, and always will be a challenging place to live, survive and prosper. Amazingly though, we've overcome the insurmountable, and have thrived as a species, now some 6 billion strong worldwide.

We're here today because the human species has been successful at adaptation. We've done a better job than many other animals, mammals, birds and prehistoric beasts at surviving the un-survivable, and will not likely find ourselves on the endangered species list any time soon. Michael Crichton did make a good point when he said,

> I suspect people of 2100 will be much richer than we are, consume more energy, have a smaller global population, and enjoy more wilderness than we have today. I don't think we have to worry about them.

Although I see the fingerprints of human-induced global warming, and I don't necessarily agree with everything Crichton said, I do feel that we will persevere. But will the people of 2100 be richer as Crichton proposes? Will our future generations consume more energy yet enjoy more wilderness? Personally, I doubt it. But who am I to say. I am merely a mortal man whose time on this Earth is but a passing flicker from a simple spark thrown from much bigger flames fueling the fire that sustains our planet and human race.

In the next chapter, I'll be discussing what the future may hold for us regarding the impacts bestowed from a warming world. Before doing so however, I'd like to first discuss the harbingers and bellwethers that, if found to come to fruition, will be ringing not only the global warming alarm bell, but also a loud and blaring siren telling us all that action must be taken. If these bellwethers do manifest themselves, our actions may not be quick enough to avert disaster; instead, we may only have time to deal with the problems, and learn how to manage a new and transformed world.

How Thin is Frog Skin?

As a young boy growing up in rural America, I became acquainted with some of our world's amphibious and reptile species early on in life. I don't know of a young boy growing up near streams, rivers or ponds who hasn't accepted the challenge of capturing a creature that can either jump, leap, hop or squirm away so quickly as to avoid capture. I was no different. I loved catching frogs, toads, tadpoles and garter snakes, all of which were never well accepted as offerings to my Mother, yet provided me with enjoyment for hours on end.

I always have been fascinated with the smallest of God's little creatures—simple beings that can be found residing near the banks of any nearby pond or creek. Frogs were among the most curious of creatures though, as these interesting critters were not quite land animal, yet not quite fish either; they were something in between, something I'd later learn was known as an amphibian. But when I was just a young boy, all I knew was that these things were called frogs, and boy were they fun to catch. Once captive however, these seemingly strong creatures, that could endure both land and water, were actually quite fragile, and never lasted long as a pet. I learned at an early age that amphibians are not as tough as they appear, and were indeed some of the most fragile varmints on our planet. As I got older, I learned also that their fragility served as Earth's barometer of health.

Frogs survive so well on land and water because of their unique skin. Their flesh is not waterproof however, and is permeable, soaking up water, oxygen, and carbon dioxide as well. When underwater, frogs can use their skin in a similar manner as fish use their gills, allowing them to breathe by extracting oxygen from the water. Yet when returning to land, frogs can breathe like any other land animal by using their lungs. While the unique permeability of frog skin allows these amphibians to go from land to water without reservation and live the best of both worlds, it leaves frogs vulnerable to changes in the environment—changes that can occur in a warming world.

Since the 1980s, there has been a dramatic decline in frog populations worldwide. Much of this is not related to global warming; instead, small variations in the balance of nature can decimate the delicate amphibians around the globe. An international convention of amphibian biologists in 2004 has indicated that over a third of the world's amphibian species are threatened and more than 120 species have likely become extinct since the 1980s.[180] Back in the 1980s, many scientists were slow to react to research that revealed declining amphibian populations, and as with the skeptical global warming view, many thought this was just a natural cycle of nature. Over time however, most scientists have viewed the reduction in amphibious numbers as real, and a severe threat to our world's biodiversity.

There are many reasons why frog populations have declined in the past couple of decades including habitat modification from encroaching human development, the introduction of foreign species such as fish brought into streams, ponds and the like that consume more than their fair share of tadpoles, and pollution has done a good job at poisoning the frogs as well. While the infiltration of foreign species and the reduction in frog-inhabited wetlands is bad news for our world's amphibians, the thin, permeable skin of frogs is serving as a kind of canary-in-a-coalmine, revealing harbingers of human-induced pollution and global warming as well.

While pollutants can take many years to manifest their adverse effects on most species, amphibians reside precariously close to the frontline since their delicate skin soaks up contaminants rather quickly. Chemical pollutants, such as insecticides, have had devastating effects on frogs over recent years, causing developmental deformities such as extra limbs and malformed eyes, alterations in their central nervous systems, and disruptions in their ability to produce and secrete hormones. The effects on frogs from changes in the environment have dramatic implications. There aren't many other species out there today growing extra limbs or eyeballs. Even fish, which we know today may contain higher concentrations of mercury and PCBs from years of human-induced pollution, aren't being mutated in the way that frogs are. Frogs are indeed some of the most delicate creatures on our planet. Their vulnerability to the slightest of environmental changes provides scientists with a view of the future.

Of particular interest now, in terms of a warming world, is not necessarily the pollutants creating extra arms and legs on frogs; instead, it's disease. Since the 1980s, when scientists started seeing a dramatic reduction in amphibian populations around the world, a disease-causing chytrid fungus with a tongue-twister of a name known as *Batrachochytrium dendrobatidis* became more prolific, and has been making its way into the amphibian world—and in a very bad way.

Chytrid fungi are typically not that harmful to most species of animals on the planet, but it's bad news for frogs. Frogs infected by chytrid fungus-type diseases tend to grow skin lesions and acquire a thickening of the skin known as hyperkeratosis. As frogs become infected with the chytrid fungus and their skin begins to thicken and grow lesions, it makes it impossible for the poor amphibians to breathe through their permeable skin, which eventually causes them to die.

A report published in the journal *Nature*[181] in early 2006 discussed how this chytrid fungus has become more productive due to global warming, thus allowing the fungus to spread far and wide to new regions. As it turns out, global warming has increased the evaporation in the tropical mountains of South America, which in turn has promoted cloud formation. This increase in cloud cover likely decreased daytime temperatures in these tropical regions by blocking out sunlight, while at the same time acting like a blanket to insulate the surrounding land, thus keeping nighttime temperatures higher than normal.

The study's lead author, Dr. J. Alan Pounds, resident scientist from the Monteverde Cloud Forest Preserve and head of the Golden Toad Laboratory for Conservation in Costa Rica, believes that the slight fluctuations in temperature have brought on ideal conditions for the chytrid fungus to spread. He states with "very high confidence" that large scale warming is a key factor in the decline of many amphibian species in South America, and in his report he iterates his pro-global warming stance that this is a bellwether for our planet by saying:

> We propose that temperatures at many highland localities are shifting towards the growth optimum [for the chytrid fungus], thus encouraging outbreaks. With climate change promoting infectious disease and eroding biodiversity, the urgency of reducing greenhouse-gas concentrations is now undeniable.

As with everything that's published on the topic of global warming, Pound's work has had its critics and skepticism. Some researchers feel that Pound's findings are just coincidental, that only a circumstantial correlation is shown, and that Pound's report fails to show the binding correlation of cause and effect. Nevertheless, Pound's paper was peer-reviewed and published, and it does have its supporters as well.

In any case, we do know that frog fragility serves as an ideal indicator of our planet's health. While debaters bicker on the specifics of global warming and the effects on amphibian species, we can trust what happens with the world's frogs as an early warning detection system for a warming world. But to tie cause to the effect, and see if human-induced greenhouse gas emissions are to blame, we'll need to turn our attention to the skies.

GASEOUS MAXIMUS

In 1973, Nobel laureate and chemist Frank Sherwood Rowland, along with Mario Molina, who would later be awarded the Nobel Prize for Chemistry in 1995, found something disturbing about our atmosphere while conducting research at the University of California in Irvine. They found that organic compounds known as Chlorofluorocarbons (or just CFC for short)—ubiquitously found as aerosol propellants in such things as hairspray and deodorant, and as Freon in refrigerators and air conditioners—could have devastating effects on our atmosphere, and eventually our planet. Rowland and Molina discovered that these CFCs did a great job of depleting the Earth's ozone layer, a part of Earth's stratosphere that works as a natural sunscreen. Without the ozone layer, we'd be in trouble, as the Sun's powerful radiation would penetrate toward our world's surface causing dramatic effects on the health and wellbeing of every living thing on the planet. With every squirt of hairspray and each dose of deodorant, combined with the manufacture and disposal of refrigerators and air conditioners around the globe, these CFCs were being released in exuberant abundance in the 1970s and 1980s, traveling then skyward to eat away at our thin yet precious ozone layer. If this wasn't stopped, and CFC emissions were to continue, the world would inevitably have headed toward disaster.

By 1976, the U.S. National Academy of Science released a report confirming that CFCs were in fact depleting the ozone layer, and that action needed to be taken. By 1987, action was taken with the construction of a worldwide treaty known as the Montreal Protocol. This treaty now has over 189 countries committed to the elimination of ozone-depleting CFC manufacture and production worldwide, and it saved our skin—literally.

Had our ozone continued its deadly demise of continued depletion, spending time outdoors would have inevitably required plenty of protection from the searing rays of the Sun. Luckily though, CFCs were stopped, and things have changed for the better in recent years as a result. Our ozone layer is recuperating (slowly) and the abundance of ozone-depleting gases has been reduced since they peaked in 1994.[182] On the downside, non ozone-depleting gases that have replaced CFCs are now being pumped into the atmosphere at higher levels than before, and, while not deteriorating the ozone layer, they are working as greenhouse gases.

Nevertheless, history does tell us that if we keep an eye on the sky, we can not only peer into the future, but also take steps to change it for the better. While we still have to deal with the greenhouse gas substitutions that replaced the ozone-depleting CFCs, we can rest easy knowing that our ozone layer is making a comeback. We can also stand confident knowing that when the world works together, we can improve our planet's future. But that was then, and this is now. While the theories that showed a link between human-induced CFCs and ozone-depletion gained worldwide acceptance nearly twenty years ago, the battle over human-induced greenhouse emissions and its fate to our world rages on.

As greenhouse gas emissions rise, a plethora of detrimental outcomes looms on the horizon, many of which will be discussed in the remaining sections of this chapter and the next. These projections are premised on the greenhouse effect induced by the accumulation of gases like carbon dioxide and methane. Similar to the CFC issue of long ago, by knowing the cause, we can see the effect. Moreover, once the cause is found, steps can be taken to remedy the situation by going directly to the source—cutting back on greenhouse gas emissions.

Improvement however will likely come late. Carbon dioxide has a very long shelf life that can last up to 200 years.[183] Even if we were to stop emitting carbon dioxide now by shutting down all of the coal-burning power plants around the world and start riding bicycles to work, the Earth would still need a very long recovery. An extremely long period would be required in which to process the excessive amounts of accumulated carbon dioxide through sinks like the oceans, plants and trees, to reduce greenhouse gas concentrations to manageable levels.

But what exactly is a *manageable* level? This is an excellent point similar to one raised by Senator Inhofe when he posed the question regarding just how hot or cold our planet should be. We know that the Earth has undergone drastic temperature swings, and an *ideal* temperature is left up to opinion based on how one views the world and how it should look and behave. Greenhouse gases are different however, since we know that we're now at all time highs, and that modern humans have never before seen the concentrations of greenhouse gases that we're experiencing now, or will in the future. Since carbon dioxide concentrations were around 280 parts per million (ppm) before the Industrial Revolution, we can set a stake in the sand and use this level as a marker or a goal. This goal is unreasonable however, since it would take hundreds if not thousands of years to attain such levels, even if we somehow were to shut off the human-induced carbon dioxide pump right now.

Currently, we're dealing with a concentration of about 372 ppm of carbon dioxide in the atmosphere today, for an increase well over 30% of what we had right before the Industrial Revolution got underway. This also beats the record set back some 323,000 years ago when carbon dioxide concentrations got up to 300 ppm. According to the IPCC projections, carbon dioxide concentrations will undoubtedly continued to rise, and will likely increase to 540 to 970 ppm by the year 2100 for an increase of 90 to 250% above what we had in the 18th century. While a carbon dioxide level of only 280 ppm is unlikely to be attained at any time in the near future, the scenarios that the IPCC came up with in their third assessment report indicate that things around the globe could stabilize if we kept carbon dioxide concentrations at least lower than they are today. To achieve such a lofty goal won't be easy though.

The IPCC, in its estimates, calculates that to get carbon dioxide emissions under control and return concentrations to a level where global warming can be stabilized, we'd need to drop our emissions to 1990 levels within the next few decades, and continue to steadily decrease our emissions thereafter for centuries—not years, not decades…centuries. While this would be a monumental task in itself, nature may add to the dilemma if we don't address this problem soon.

Permafrost—that solidly frozen land in Arctic regions—is melting away with a burgeoning problem. It's estimated that nearly 25% of the Northern Hemisphere is covered in this icy ground, and that billions of tons of organic matter lie locked safely away under this vast expanse of frozen tundra.[184] As this land thaws, the organic matter will have a chance to release its greenhouse gas byproducts. This would occur through a process known as *thermokarst*, where rising air temperatures melt the normally frozen and flat permafrost into a series of depressions and mounds where water pools up, which accelerates the thawing process until the permafrost is completely melted. During this process, both methane and carbon dioxide byproducts from the permafrost's organic matter get released into the atmosphere. In Siberia, where permafrost is melting at an alarming rate, it's estimated that 70 billion tons of methane are stored within the frozen ground and that in the past 40 years, air temperatures in the region have increased by 3°C (about 5°F).[185] This will likely cause a feedback loop where rising air temperatures melt permafrost, thus releasing greenhouse gases, which then warm the planet to exacerbate the melting process.

Feedback loops in nature like the permafrost issue, combined with human-induced emissions will undoubtedly make any greenhouse gas reduction plan a difficult task—like catching a fast moving freight train that refuses to stop at the station. Achieving the 1990 emission levels recommended by the IPCC, and continuing to reduce our emissions for centuries will not be easy and it may seem hopeless. Yet many thought it would be impossible to eliminate CFCs back in the 1980s, and that it too would take so long that a "what's the use" mentality became commonplace. We do know from the historic lesson on CFC reduction though that we can indeed take steps to remedy a long-term problem, and there are things we can do, which I'll discuss in chapter 8. The time to act though is always sooner than later. The longer we wait to reduce our own emissions gives feedback loops in nature more of a chance to accelerate, which could lead us to a precarious precipice where a tipping point is pushed by butterfly-effect-like circumstances.

As the debate over human-induced global warming carries on, each day, we continue to emit more and more carbon dioxide and other greenhouse gases into the atmosphere. At some point, we will learn that either humankind is pumping too much greenhouse gas into the sky, or that the world somehow will be able to handle it. With the mounting evidence pointing to human-induced greenhouse gases being on the rise and not likely to go away on their own, time should be of the essence to take steps to curb emissions and work toward a worldwide reduction. Yet skeptics choose to debate the issues to the point where they cast enough serious doubt on human-induced climate change that lawmakers, relying on the voice of science for sound information, hear conflicting views. This slows the progress in finding solutions as policymakers sort through the white noise and chatter coming in from both sides of the global warming controversy.

Sadly, policy is often slow to come until disaster strikes. Stoplights are put into intersections to replace stop signs more often than not only after a speeding motorist mows someone down. The theory that danger looms on a busy road is typically not enough to persuade action from the authorities. As a result, many suffer needlessly as lawmakers look at only hindsight and doubted projections that are hotly debated. Human-induced global warming has a long roadmap of research, and the intersections of intervention are only now being designed. Enforcement to control these roads that lead into our future, like nearly everything else that comes from those in charge, will undoubtedly be slow in coming.

If carbon dioxide, the biggest contributor to greenhouse gases is found to be in decline in the future, this will be good news. If not, then we'll be faced with a harbinger of further effects from a warming world.

TEMPERATURE TRENDS

On the evening of March 10, 2006, something bizarre happened in the typically sunny region of Southern California. As a giant storm raced down the California coast, it brought with it not only rain from the Pacific, but cold air from the Arctic as well. Hail, an unusual sight near coastal communities near Los Angeles, pelted the streets and freeways, leaving a dangerous sheet of ice on the roadways. As winds picked up and howled from the north, temperatures continued to drop, with nighttime temperatures around LAX reaching no more than barely 40 degrees—a substantial drop from the norms that hover around the 50s at night in March and near 70 during the day. In the morning, once the bulk of the storm had passed, I peered out of my bedroom window at the hillsides near my home and saw something I had rarely ever witnessed before on those low elevation peaks by my house: snow.

In talking with a friend of mine that day, commenting on how the blustery daytime highs of only 52 degrees had us bundled up, looking upon the snow capped hills surrounding my humble abode, I felt a little foolish when mentioning that I was working on a book on global warming. Warming? Where? Southern California, only a couple weeks away from the first day of spring, had endured a nasty cold snap, leaving most Southern Californians unprepared in a land where dressing formal means not wearing shorts and flip-flops. With air temps more reminiscent of Idaho and not California, the idea of a warming world seemed a bit absurd.

The cold snap didn't last for long though, and temperatures quickly rebounded back to the usual high 60s within a few days. If I were to be looking for a harbinger of global warming in temperatures, it sure didn't manifest itself on that anomalously chilly day in March 2006. Mother Nature has her mood swings from time to time, and like that storm in March, bizarre events do take place naturally. Moreover though, my view of temperature change, whether my observations were from the California snowfall in March, or a balmy day in the summer would be short sighted. My one solitary thermometer sitting at a single point on the planet doesn't tell the whole story. It's similar to the parable of the stork and the frog, where the frog sees the entire world as a pond—he eats, breaths and breeds in his isolated myopic microcosm consisting on just one small body of water. Yet the stork, flying overhead sees many ponds with many frogs, viewing the world much differently. Perspective means everything when talking about our world. And when gauging temperature, the more ponds and places we can observe, the better our view of the world will be.

Since the late 1800s, our average global temperature has risen by about 0.6°C (around 1°F). As greenhouse gas concentrations increase, this temperature is expected to climb. The IPCC estimates that we will likely see another 1.5 to 4.5°C increase by the year 2100—in Fahrenheit that's about 2 to 8 degrees.[186] This is a wide margin, since each tick of the mercury—as shown in Mother Nature's Goldilocks Complex—has dramatic butterfly-effect-like ramifications. Yet nailing down just how much the temperature will rise is difficult.

One of the variables that will determine just how much the Earth will warm is the concentration of greenhouse gases in the atmosphere. While we know that greenhouse gases have increased in recent times and will continue to increase in the future, we don't know exactly just how much more carbon dioxide, methane, and other such elements will be emitted by our future generations. If the world's population never changed, and we kept on going as we are now, then we could narrow in on a more exact number. But government policies that could govern the use of fossil fuels in the future, changes in demographics as people move about (possibly from drought and other conditions that inspire people to move elsewhere in a warming world) and social and economic development all can play a part in how much greenhouse gas is pumped into the air each year. If economies grow, and we continue to burn fossil fuel to provide energy to sustain this growth—for more cars, bigger houses, etc.—then more greenhouse gases would be emitted compared to a scenario where economies wouldn't thrive as well, and subsequently wouldn't require increased amounts of energy. All of these factors will determine how much greenhouse gas will be pumped into the air over time, and in turn, how warm things will get in the future. But these are all unknowns at this point, thus leaving a broad range of possibilities from only 1.5 to as high as 4.5°C by 2100.

Even at the low end of this projection at 1.5°C, the Earth will continue to change. Bear in mind however, that these temperatures refer to global averages and not all regions will be affected equally. Since land warms faster and hence more than the oceans, it will take a while for ocean temperatures to increase in coming years, yet they will likely stay cooler than the land itself. There should however be a more rapid warming at high latitudes. This would be caused from an ice-Sun feedback loop where ice that would normally reflect about 90% of the Sun's rays would be gone due to melting, allowing more sunlight to penetrate land that was once covered by glaciers, and seas once blanketed by ice. Since land only reflects about 20% of the Sun's rays, and ocean water only around 10%, the absence of Sun-reflecting ice would create a warmer surface in at least the Arctic, North Atlantic, and North America. This would then accelerate ice melting, which would exacerbate surface warming even further.[187]

Conversely though, climate models indicate that the North Atlantic region will see smaller increases in surface air temperature. Thus, judging the variations in temperature from one region to the next will undoubtedly vary, requiring that a global average temperature be taken, ensuring that the Earth is not just in a zero-sum cycle heating one region while cooling another. All regions would need to increase in temperature to reveal that the world as a whole is indeed warming; otherwise, we couldn't call it *global* warming. Note however that the temperatures referred to thus far (the 0.6°C since the 1800s and the 1.5 to 4.5°C by 2100) do refer to global average temperatures. This will continue to be the yardstick to use as a measurement for bellwethers in temperature fluctuations.

There is also expected to be a decrease in the difference between daytime high temperatures and nighttime lows. This is thought to be occurring in some regions of Central and South America now, which would be responsible for the proliferation of the chytrid fungi linked to frog extinctions mentioned previously. Increased cloud cover could cool daytime temperatures while insulating the Earth in many locations, keeping nighttime temperatures warmer. A number of climate models also show a general decrease in this kind of temperature variability from day to night during winter months, yet an increased variability during the summer seasons. This is bad news for sea ice and snow cover in the Northern Hemisphere, as the overall winter temperatures would be higher than normal (especially at night), which would result in increased melting.

Although temperatures will likely continue to increase over time, the IPCC does show scenarios where we would hit a maximum temperature and then level out, and hopefully, over time, begin to decrease. This settling-down scenario is based on the premise that greenhouse gas emissions will be reduced in coming years. As these gas emissions diminish, the rise in global temperature would begin to level off. The stabilization of temperatures though takes far longer than any reduction, primarily due to the long life of greenhouse gases like carbon dioxide and methane. After a reduction in carbon dioxide and methane, it would still take a hundred or more years to dissipate and recycle back into nature the surplus of these greenhouse gases that have been added to the atmosphere over so many years.

Surely, over time, as new readings of global average temperatures are calculated, the media will have new headline material to broadcast to the masses. This will serve as a solid bellwether, since temperatures, more than any other signal, will warn us of what will be coming. Debates can rage on over how much greenhouse gas is in the atmosphere, and what effects it may or may not have on temperature. What can't be disputed though are the basic laws of physics: heat something up and ice will melt, sea levels will rise, and our weather will change.

The crux for climate change is indeed temperature itself, which cannot be disputed.

ICE AND THE RISING SEAS

Every day the tides ebb and flow around our world's vast oceans. Sea levels rise, and later they recede. Yet we know that lately the overall depth of the oceans has become greater, encroaching slowly over land and changing the way of life for many people around the globe. Today, we know that during the 20th century the average sea level around the planet rose anywhere from four to eight inches, most of which has been attributed to thermal expansion from warming waters.[188] We also know that sea levels don't immediately rise when air temperatures do, since it takes a bit longer for the seas to heat up, and for thermal expansion to take effect. So in a world that's warming now, could we expect sea levels to rise even more while waiting for that lag from temperature increase to thermal expansion to take place? Moreover, much of the sea level rise we've experienced in recent times is due to thermal expansion, yet glaciers are now melting, and we have signs that Antarctica is doing the same. Will this add more water to our oceans and increase them even more? The answer is simply, Yes.

Temperature will be a bellwether that will warn us of many things, including rises in sea levels. As sea levels rise, ecosystems will be transformed, shoreline residents will be affected, and land will inevitably disappear. As mentioned earlier, our world has seen drastic swings in sea levels, even higher than what we've seen today. Before our last major ice age started some 120,000 years ago, the warm period that preceded it saw sea levels over 15 feet higher than they are now. When the ice age peaked some 18,000 years ago, glaciers and ice sheets had drawn so much water out of the world's oceans that the sea level dropped an astonishing 300 feet lower than it is today.[189] So when we talk about an increase of 4 to 8 inches, which occurred during the 20th century, it sure doesn't seem like much. Earth can take it; she's done it in the past, and she can do it again. But life is different now on planet Earth; many more people live in our world today, and many live precariously close to the sea.

It's been estimated that nearly half of humanity inhabits the coastal regions around our world.[190] Millions of years of ocean enrichment on these low-lying soils from previously higher sea levels, and erosion runoff from higher elevations, have left so many coastal areas fertile and ideal for farming, and the subsequent growth of societies. Humans, to break away from being just hunters and gatherers have always needed farmland to survive. Coastal zones provided just such a sanctuary for plentiful harvests. And now, in an age where sea levels had barely budged for hundreds if not thousands of years, people living by the water's edge will need to be aware of an increase in the depth of our oceans.

Bangladesh serves as a good example of how a miniscule rise in sea level could have dreadful effects on a civilization that's grown accustomed to living close to the sea. In this low-lying country of about 120 million people, located on a delta region bordering three major rivers, it's estimated that about 10% of the entire country's habitable land would be lost with only a foot or more rise in sea level. This land, being so close to the water's edge, would displace about 6 million people. Considering that the New Orleans catastrophe from hurricane Katrina displaced tens of thousands of people, the exodus of Bangladesh would be magnitudes greater.

Other regions are at risk from the rising seas, including the Netherlands, where 8 million people live below sea level now. While a well-engineered system of dykes and coastal dunes is maintained to keep the seawater out of populated regions, rising sea levels will require continued improvement to avoid breaches of the water systems like the 2005 disaster that happened in New Orleans. And of course, those low-lying islands in the South Pacific will be faced with disappearance as well, and coastal areas everywhere will lose beaches from erosion.

Rising sea levels are a major concern to so many people around our world, but unfortunately, there is little we can do to prevent it—we can only prepare for it, and learn to manage it as well. The IPCC points out that if greenhouse gas emissions were to stabilize now, the sea level would continue to rise for hundreds of years due to the long lag in thermal expansion and other factors such as slow oceanic circulations. After 500 years, sea level increases may only reach half of their eventual peak, which climate models say could be anywhere from 0.5 to 2.0 meters (about 1.5 to 6 feet). This is just from our recent warming trend. While that's disturbing in its own right, the IPCC also estimates that if carbon dioxide levels double or quadruple, the increase would be around 1 to 4 meters (about 3 to 12 feet). At the low end of 3 feet, Miami would be underwater, as would much of the Florida peninsula and the Gulf Coast. Kick that up a notch towards the high end of 12 feet, and things get much, much worse.

Going hand in hand with the increasing seas would be the continued melting of glaciers and other various ice sheets. According to the IPCC's third assessment report, ice sheets will continue to react to climate change for the next several thousand years, even if the climate were to become stabilized. While the hundreds of years of rising seas may send shivers down your spine, that is really only a fraction of the thousands of years in anticipated glacial decline around the world. This is exceptionally frightening when you consider that the complete melting of the Greenland ice sheet would result in an additional sea level increase of around 20 feet, and recent reports are showing that Greenland's ice is melting faster than ever anticipated.[191]

Antarctica's ice is also a major concern, since, if all of Antarctica's ice were to melt and Greenland's ice were to thaw out as well, sea levels would increase a mind-boggling 120 feet. Yet, it's estimated that most of the melting from global warming will occur in the north, sparing Antarctica the drastic melting that Greenland would endure. Still, IPCC scenarios do show that Antarctica's melting would contribute around 3 millimeters to sea level increases per year for the next thousand years or so. This projected sea level rise from Antarctic meltwater would come out to around 300 millimeters by 2100 (around 11 inches). While this seems minor, bear in mind that this estimate, coming from the IPCC third assessment report, was stated in 2001, and as I mentioned some chapters back, a study conducted from 2002-2005 showed that the ice sheet mass covering Antarctica has been rapidly melting at a rate of 36 cubic miles per year—far more than expected.[192]

Rising seas will no doubt have an impact on our way of life, leaving future generations with a different world than we have today. Although our planet has seen this type of thing before, the transitional period was always slow, allowing eventual adaptation, and not overnight responses. In our current warming world, changes are occurring rather quickly; thus, the bellwether of rising seas and melting ice should ring a sense of urgency, forcing us to quickly learn how to manage a changing coastline around the globe.

Weather Extremes

The term *climate change* is often thought of as just a warming up of our planet, with visions of melting ice and scorching heat. Earth though is a tightly woven mix of myriad weather with all spectrums of extremes being closely linked to one another. Yet in many respects, perturbations to Earth's climate system are similar to simply stubbing your toe. When you accidentally ram your foot-extended digits into something that just won't budge, your toe will respond via nerve endings transmitting a pain signal that within a split second travels up to your brain, making you wince and yell expletives at the top of your lungs. A complex system of nerves, blood, flesh and vocal chords came together to exhibit the pain felt from a localized disturbance—jamming your foot into an immoveable object that was much stronger than your tiny toe. This intricate system of the human body provided many a reaction from a simple action, from which, you were able to avoid further pain, and not repeat the mistake again. A complex system provided these means, but the initial impact and pain were just the beginning. After the stubbing incident, you'd have difficulty in using your foot. Subsequently, you'd hobble about for a bit to avoid putting pressure on your injured extremity, which would result in even more pain coming from your wounded toe. If you limped around long enough, you'd put excessive strain on your good foot and leg, but you'd exert less work on your injured side. This would then disturb the natural balance of your elaborate, humanoid system.

Within a few minutes, your toe would likely heal itself, and all would return to normal—no big deal. But if you rammed your toe hard enough into a solid object and broke one of your tiny appendages, then you would undoubtedly stagger about for an extended period, putting even more exertion on your favored leg and far less on your injured limb. The injury that affected just a mere toe would now start to affect all the joints and muscles within both of your legs. Within days, your favored leg would end up doing most of the work to transport you to and froe, thus becoming stronger while the leg from the injured toe begins to atrophy and becomes weaker as it feels it's no longer needed. Spawned from a simple butterfly-effect-like incident—stubbing your toe—your body would go through changes to compensate for weakness in one area by building strength someplace else.

Our world is quite similar, although Earth doesn't really have any toes to stub, or legs to hobble about on. Nevertheless, Earth's intricate system of weather works in a comparable, zero-sum kind of way. As one area becomes weak, another becomes stronger, thus providing equilibrium where two extremes are found. This disparate dichotomy formulates an average of neutrality. Yet seemingly insignificant incidents—similar to stubbing one's toe—can occur in a warming world to disrupt nature's balance. We've seen this in ocean oscillations and weather phenomena such as the ENSO events, where El Niños, kicked off by tiny increases in temperature, cause active hurricanes in the Eastern Pacific but not the Atlantic. Similarly—yet conversely—when a La Niña is set in motion by miniscule modifications of sea surface temperatures, the opposite is true. Earth truly lies in a balance of nature: when one area becomes adverse, another prospers. It's a ying and yang kind of thing, where we have hot or cold, good or evil, night or day, paper or plastic, and mustard or ketchup. Our world tends to work in a zero-sum state of stability, all set off from ostensibly small butterfly-effect kinds of agitations. This is especially true in a warming world and its weather: when one area becomes dry, another area can become quite wet. But as the world heats up, the trivial disturbances that set opposing forces in motion become accentuated, and the game of zero-sum rewrites the rulebook.

Overall, as greenhouse gases continue to trap the Sun's heat, *all* land areas are expected to become hotter. Earth's zero-sum game is thrown off balance when it comes to temperatures, from which everything else is linked. Since the atmosphere as a whole will contain an increased concentration of greenhouse gases, Earth's entire protective gaseous blanket becomes thicker everywhere. Although our world's layer of greenhouse gas-rich atmosphere could become thicker in some places more than others—from such things as localized cloud formations—overall, the atmosphere will do a good job at circulating greenhouse gases worldwide. This overall thickening of Earth's atmospheric blanket will result in more hot days everywhere, and heat waves could occur more often as well. Frost and cold snaps would be less frequent too, since overall we'd be warming up across the globe. The increases in temperature worldwide, although not zero-sum in their own right, would set off a zero-sum effect around the world, changing precipitation levels from place to place.

Earth's precipitation is controlled through what's known as a *hydrological cycle*. This cycle has kept our planet alive by continually recycling the Earth's supply of water, which is finite. The water we use now has been around for billions of years, just in different forms. This hydrological cycle is a bit of a chicken-and-egg paradox that spins in a loop with no real beginning or end. In the hydrological cycle, water vapor in the atmosphere condenses to form clouds, which release rain when conditions are just right. Precipitation falls to the surface of the Earth and infiltrates the soil or flows out to the ocean as runoff. Surface water from lakes, streams and oceans, as well as from plants, evaporates, which returns moisture back into the atmosphere where the cycle begins all over again.

This hydrological cycle has been going on for a long time, ever since the first drop of water came to be on our planet. Every drop of water we drink now fell as rain at sometime somewhere. It then formed clouds as it evaporated, and some of it ran into waterways while plants soaked up some of it too. People and animals drank the water, released it as waste, which ran off somewhere and eventually evaporated as well, thus reentering the hydrological cycle to eventually be formed into clouds, rain, etc. The water in our oceans, streams and lakes has been moving through the hydrological cycle for eons, consumed by not just humans and plants, but other creatures of the Earth as well. In fact, you could be drinking recycled dinosaur pee right now. As bizarre as this may sound, the hydrological cycle is real—our water just doesn't get created and destroyed everyday; Earth only has a certain amount of water and it's constantly being recycled. This recycling process of the hydrological cycle is what makes precipitation around our world a bit of a zero-sum game, especially as the Earth heats up.

As arid regions become hotter, more of the remaining water that's left in the ground and surrounding water supplies evaporates, intensifying the dryness of drought-prone regions. Warming of waters elsewhere around the globe also accelerates evaporation, thus all regions of the planet release more water into the atmosphere. This excessive release of moisture into the atmosphere adds fuel to the hydrological cycle, thus accentuating its effects. While the drought-stricken regions will continue to be dry in a warming world as water is sucked out of the nearly bone-dry soil, the enhanced hydrological cycle will dump more rain to regions that normally experience decent amounts of precipitation. This then increases the risks of flooding with higher than normal rainfalls occurring in most non-drought regions. Thus, areas that see rain on a regular basis today will—in a warming world—see even more rain with possibilities of flooding, while regions that pray for rain will likely see further stresses from drought.

With more storms bringing added rainfall to wet regions, there is increased likelihood for other extreme weather events such as thunderstorms, hail and tornadoes. Since more rain-events could occur and last longer, the odds become favorable for extreme events related to rainmaking systems. Yet, it should be noted that many climate models disagree on just how extreme tornadoes, hail and thunderstorms might become. Nevertheless, climate models do agree on the augmented hydrological cycle with more rain to normally wet areas, and dryer conditions to regions already in the grips of drought.

There is still quite a bit of uncertainty in future predictions of extreme weather in a warming world. While climate models calculate rising temperatures, so many variables get thrown into the mix such as cloud cover, which could cool things down during the day, heat things at night, and add rain into the hydrological cycle. While we do know that temps will warm and the hydrological cycle will be agitated, models can't seem to agree on what will happen with big weather cycles like El Niño. There is a consensus in model simulations however, that show a small increase in El Niño events over the next 100 years. Nevertheless, the best bellwether for weather will undoubtedly come from things directly related to the hydrological cycle: increased rain in normally wet regions that will likely lead to flooding, and dryer conditions in drought-prone regions that can lead to famine and fire.

Increased rainfall and drought intensification are inevitable bellwethers. Yet more research needs to be conducted in coming years to understand how cyclic phenomena will be affected in a warming world. After all, we know from recent history that amplifications of storms and the cycles that enhance them need our undivided attention.

THC AND HOPPED-UP HURRICANES

Life in Florida and along the U.S. Gulf Coast is a gamble from the beginning of June through the end of November—the official hurricane season in the Atlantic. Every year since records began back in 1851, somewhere along the U.S. coast a tropical storm has made landfall. Not once was the U.S. coastline spared a direct hit in any one given year. From 1851 to 2004, 273 hurricanes have made landfall in the U.S., hitting the coastline somewhere between Texas and Maine. Florida has seen the most hurricanes, comprising nearly half of the total 273-hurricane tally with 110 storms coming ashore since 1851. Louisiana ranks up there as well with 49 hurricane impacts, and Texas has seen 59 cyclones as well. Hurricanes are part of life for people living in these storm-prone regions. Knowing this, local governments have enforced building codes and worked together on emergency response and evacuation plans to avoid disaster. Yet no amount of planning can completely avert a cataclysmic event from a storm whipping up winds well in excess of 150 mph.

Emergency crews do what they can, residents board up houses, and some flee the area entirely once a storm bears down on the U.S. coastline. Forecasters have done a good job at perfecting the science of hurricane tracking over the years, yet these rogue storms continue to fool the fastest of computer models, and at the last minute, these unruly cyclones can shift course with unexpected devastation. The unpredictability of hurricanes complicates the coexistence of humankind with nature, requiring civilizations to dance lightly in nature's ballet while allowing hurricanes to take the lead and call the shots. We can't stop them, but we can learn to deal with them.

The call to evacuate from the path of a hurricane is a costly one, running well into the millions of dollars in lost revenue from businesses that shut down while residents board-up and flee, coupled with the increased transportation costs to pull people out of harm's way. Forecasting a storm's trajectory incorrectly can result in unnecessary lost income for many and can cost local governments dearly as well. This however, pales in comparison to the costs to rebuild storm-torn regions, which can easily run into the tens of billions of dollars. The biggest price however, that of human life, outweighs all other monetary repercussions, punctuating the seriousness of living in the firing line of nature's fury.

When realizing the costs of property and human life lost to the ravages of nature's most powerful storms, it's not hard to comprehend why Dr. Kevin E. Trenberth, much to the chagrin of Chris Landsea, stood before the American people and said what he did. Ternberth surely understood the gravity of the matter, which justifies his attempt to explain why the U.S. had endured such a devastating hurricane season and forewarn all that we may see more of the same very soon. The cost of hurricanes, both personally and financially, is so high that no one should take these vicious storms lightly—not even the most ardent of skeptics among us. These storms are far too serious to ignore, and although Landsea may be justified in his concerns of others crying wolf, urgency is an issue when it comes to hurricanes—we need answers, and we need them now. One can only defend against a vicious attack once they know their enemy inside and out. Still, there is very much that we don't know about hurricanes and their links to a warming world. We're familiar with the signposts of hurricane formation and harbingers of climate change, yet even these have proven to elude us at times.

The jury is still out on whether we can use hurricanes as reliable and consistent bellwethers of global warming. Although studies conducted recently (mentioned in chapter 5) have started to point towards global warming as a cause for recent hurricane intensification, more studies will need to be conducted to find conclusive answers. Nevertheless, we do know that warmer waters will provide hurricanes the fuel they need to grow and become stronger, and that conditions as of late have been favorable for reduced wind shear and hurricane transport to the U.S. coastline, which may or may not be linked to global warming. While many signs do point to the enhancement of hurricane-generating elements in recent years, and theory could point us in the direction of global warming, further research and observation will be paramount over the next few years.

While watching for the bellwethers that can accentuate hurricane formation and intensity—such as warming waters, reduced wind shear, etc.—there is one factor that we need to keep an eye on that can affect more than just hurricanes. One of the biggest contributors to hurricane intensification and to worldwide climate as well, is the long-lived cycle in our oceans that I discussed back in chapter 3: the thermohaline circulation (THC).

The THC may have sped up in recent years, helping to warm the waters in the Atlantic and bring warm water to the Arctic as well. This would not only provide more warm-water fuel for hurricanes; it could also result in further melting of sea ice, which could lead to an eventual slowing of the THC. Over time, the THC will completely shut down—it's done that before, and it will inevitably happen again. When (not if) the THC shuts down again, our world will change. How soon the THC slows or shuts down will be one of the brightest beacons alerting us that drastic changes loom on the horizon.

Most climate models do show weakening of the THC in the future. Since global temperatures are on the rise, freshwater will continue to melt from glaciers and Arctic sea ice, which will reduce salinity. As mentioned earlier in chapter 3, this reduction in salt-water density will rob the THC conveyor from the fuel it needs to continue flowing its deep-water global currents. As more freshwater is added to the mix over the coming years, the THC will continue to slow down, until one day it will just stop flowing all together, just as it did many millennia ago.

A slowing—or complete halt—of the THC would result in a reduction of heat transported to the high latitudes of the Northern Hemisphere. As less heat is drawn towards the Arctic, sea ice will have a chance to reform, but that will take a very long time, perhaps a millennium or more. As less heat is drawn northward from the reduced speed of the THC, less warm water is pulled through the waters off the coast of Europe, which could result in cooling in England, Ireland, Scotland and France. In a world not affected by global warming, a slower THC could inevitably make Europe a bit chillier, similar to what happened in the last ice age. In a warming world however, things are a bit different.

Climate models show that land temperatures from global warming would still be higher than normal over Europe from an increased greenhouse effect; thus, offsetting any cooling from a slow or deceased THC. The current projections from the best climate models don't seem to think we'll see a complete shut down of the THC in at least the next 100 years, although it will likely start to slow down throughout this century. Beyond 2100 however, the THC could completely shut down if global warming continues on its current trend. Worst-case scenarios show that the THC could not only shut down, but also have difficulty restarting again for quite some time. This would have an interesting effect on a warming world.

The THC stopped moving around 12,800 years ago during the Younger Dryas period. With less heat transported to northern regions, ice began to form, and the Northern Hemisphere became blanketed with ice. Over time, the THC picked back up again as salt, released from the freezing Arctic ice began to sink and revved up the THC conveyor system. It's another one of Mother Nature's miraculous wonders to keep things in check in a cyclic, zero-sum world. Yet back in the grips of our last ice age, greenhouse gas concentrations were far lower than they are now. Back when Wooly Mammoths still roamed the Earth, the perturbations to the THC were quite natural with slow progressions. With increasing greenhouse gases heating things up like never before, things are much different now than right before or during the Younger Dryas.

If the THC slows or stops now, what exactly will happen? With things so different now than they were some 12,800 years ago, it's really hard to tell. Will we enter another ice age again? Or will land temperatures from greenhouse warming be too great to allow ice to form? And if ocean water temperatures remain warm from the greenhouse effect, will hurricanes be intensified, or will they wane from a weaker THC? No one is really quite sure, but scientists are very concerned nonetheless.

Climate change science is riddled with questions like these—questions that are not easy to answer, and carry an abundance of uncertainty. Looking at hurricanes and the THC, we may see harbingers of a warming world. Should hurricane ferocity and frequency diminish, this would be good news for all of us. If the THC stabilizes, this would be even better news. If either one of these bellwethers continues on their current path however, then the sense of urgency is further escalated, and let no one stand in the way of finding the truth—there's just too much at stake.

7

The Cloudy Crystal Ball

○ ○

How long can men thrive between walls of brick, walking on asphalt pavements, breathing the fumes of coal and of oil, growing, working, dying, with hardly a thought of wind, and sky, and fields of grain, seeing only machine-made beauty, the mineral-like quality of life?
—Charles Lindbergh

If I could look into the future, I'd be a very rich man. Yet I can't, so I'm not. No one truly knows what lies ahead of us; all we can do is speculate and prepare for various scenarios. I plan on growing old, so I save some money for retirement, or for emergency purposes. Will I actually grow old to spend it? Will I endure an accident and require the money to pay for bills and expenses before then? No one can tell. Yet knowing various possible scenarios of my future, I can draw a blueprint of possibilities, allowing me to plan for each possible outcome. The uncertainties of science are very similar.

Knowing the probability of each outcome for my personal and financial future, I can put emphasis on particular scenarios, giving high probability outcomes more attention than those that have a lower chance of coming to fruition. Since there is a chance I could die tomorrow, one could argue that I should spend my savings now and live for the day—*carpe diem*—I may not be here when the Sun comes up in the morning. I'd weigh that as a fairly low probability scenario, and rank my older age a bit higher. Thus, to prepare for the high probability scenario (retirement) I will continue to sock away some change for a later day. Yet there still looms the low probability scenario (tragic, early fatality), which even though holding a low likelihood, still is not impossible; therefore, it cannot be ignored. To prepare for the less likely scenario, I'll make sure I don't horde *all* of my money, but instead will enjoy some of it now. Nevertheless, this less likely scenario will take a backseat compared to my higher probability scenario of living well into my golden years; thus, I continue to stash some cash when I can.

Balancing the scenarios of our future and placing a weight of probability on each scenario is about the best anyone can do when looking into the unknown. When looking toward the future of our planet enduring the heat of greenhouse gas-induced global warming, there are myriad uncertainties. Skeptics use these uncertainties as ammo to constantly bombard any new study that points to a truly warming world with catastrophic outcomes. Perhaps its denial—who knows, I can't read minds either (I'd be exceptionally rich if I could do that *and* see into the future). Yet each uncertainty has a weight, in the same way that all scenarios do. Some things are just so probable that you'd feel comfortable betting the bank on them; yet, since the future will reveal the true outcome, it's very difficult to say with 100% certainty just exactly what all of us can expect. But we can get pretty darn close, weigh the odds, and prepare accordingly.

I mentioned a couple chapters ago that the IPCC uses some special wording to place particular weight on scenarios, including *virtually certain* (a 99% chance), *very likely* (90-99% chance) *likely* (66-90% chance), *medium likelihood* (33-66% chance), *unlikely* (10-33%), *very unlikely* (1-10%) and *exceptionally unlikely* (less than a 1% chance). While this may seem like a great way to cover their butts just in case something doesn't turn out as planned, it is intended to show the weight of probability, thus allowing policymakers, lawmakers, scientists, engineers, and everyone around the planet for that matter, to make informed decisions based on the laws of probability. Much in the same manner that each of us can prepare for our personal futures based on the likelihood of various scenarios, the outcomes for our future warming world are based on probabilities as well. It's up to each of us to interpret the probabilities, and plan ahead based on how much weight we feel should be placed on each scenario.

While we've looked at bellwethers that can warn us of a warming world, I haven't yet discussed the details of how these things may affect life around the planet in the coming years. In this chapter, I'll be covering some of these scenarios for our future, and the dangers surrounding them. There are indeed many possible outcomes in a warming world—some that are highly likely, and some that are quite unlikely. Since so much is unknown, and a plethora of possibilities exist in the cloudy crystal ball of uncertainty, I'll stick primarily to the higher probability scenarios, and things that are more *likely* to occur.

COASTAL QUANDARIES

As our planet heats up, many things will change. Artists will paint pictures depicting new landscapes, especially those along the coastal regions of planet Earth. Like time capsules, paintings today will display a world that was, but will likely not be seen the same way again, at least not for hundreds or thousands of years. After all, even if global warming were to be subdued, thermal expansion of the world's oceans will continue for many years and sea levels will continue to rise.

As I mentioned in the previous chapter, if greenhouse gas emissions were to stabilize today, sea levels would continue rising for hundreds of years. After 500 years, sea levels could easily increase by 1.5 to 6 feet, depending on the amount of greenhouse gas that's pumped into the air. If we were to quadruple our greenhouse gas emissions, sea levels could rise up to 12 feet. This is from just thermal expansion alone. Melting ice just adds to this problem. If all the glaciers *outside* of Greenland and Antarctica were to melt, sea levels would rise about 20 inches.[193] And yet if Greenland glaciers were to melt—which is happening right now—we'd have an additional sea level increase of around 20 feet.[194] If Antarctica's ice melted, then we'd all be running for higher ground as sea levels would increase to an unthinkable 120 feet.

Sea levels are rising now with a four to eight inch increase during the 20th century, most of which came from thermal expansion. This trend will continue, and there is very little we can do about it, since even an immediate halt to greenhouse gas emissions will not stop the process of thermal expansion that's currently under way. Eventually we may be able to get this under control, perhaps hundreds of years from now, but until then, coastal protection will be paramount.

While an inch here and an inch there may not seem like much in the way of rising seas, the slowly rising ocean waters bring on a deceptive form of creeping normalcy that will no doubt lead to landscape amnesia. When looking at a coastal community, one only needs to see how slightly high above sea level its residents' homes, streets, shops, docks, and everything else built by humankind, is located. Miami Beach, Florida has a peak elevation of only four feet, with much of the city located no higher than three feet.[195] A one-meter rise in sea level would nearly inundate the entire city. That one-meter rise may take hundreds of years from thermal expansion, but as glaciers melt, the rise will be expedited. Even just a small rise though could have devastating effects.

According to studies conducted by the U.S. Environmental Protection Agency (EPA)[196], it's been found that sea levels are rising more rapidly along the U.S. coastlines than other regions of the world. Studies by the EPA have estimated that the Gulf and Atlantic coasts could see another one-foot rise in sea level as early as

2050. As the sea level rises each year, beaches will continue to shrink and erode as the sea inches its way up the banks of our shores. As the water becomes higher along our shorelines, storm surge and pounding waves will eat away at beaches more and more, eroding the sand and land to greater heights than just the rise in sea level alone. This will have detrimental impacts on recreational areas that count on these beaches to lure in tourism revenue—once the beaches are gone or deteriorated, the region loses its luster. But this is also a threat to anyone living near the coast.

Many ocean shores right now along the U.S. coastline are eroding at a rate of one to four feet per year. This rate will likely continue to occur in coming years as sea levels rise from inevitable thermal expansion and the threat of meltwater from warming glaciers. Coastal engineers estimate that a mere one-foot rise in sea level will cause beaches to erode up to a foot along the shores extending from New England to Maryland, while beaches along the Carolinas erode up to four feet. While this is bad news for the eastern seaboard, this is small compared to the ten feet of erosion that could occur from a one-foot rise in sea level around the Florida coast, which could threaten a great deal of currently inhabited real estate.

Because many U.S. recreational beaches are less than 100 feet wide at high tide, that small one-foot rise in sea level would jeopardize homes in these areas. A study by the Federal Emergency Management Agency (FEMA)[197] in 2000 estimated that about 25% of all buildings within 500 feet of the U.S. coastline would be taken away by erosion in the next 60 years. Costs to U.S. homeowners from this erosion would average more than a half billion dollars per year, and additional development in high erosion areas will lead to higher losses. To protect the beaches becomes even trickier, and not always feasible. Along many sandy beaches, property owners often erect seawalls to halt or slow down erosion. Although these seawalls protect the houses, they can eliminate beaches, particularly bay beaches, which are usually less than ten feet wide, leaving the erected house as oceanfront and not beachfront property over time.

Flooding is practically inevitable from the rising seas. When judging a floodplain, engineers and flood management organizations refer to the capacity of a floodplain to withstand storms by how often a storm of a great magnitude occurs. Every year, regions get hit by storms, but every say, 15 years, a big storm comes through, and every 100 years an even bigger storm can tear through the same region. To prepare for the future, flood management organizations calculate flooding potential based on a 15-year storm (stronger than the average storm) and 100-year storms (the rogue storm with extreme ferocity that hits once a century). Planning for these kinds of storms will change as time marches on. As sea levels rise, the storms that plow into a coast can become more brutal as the seas reside closer to land, and warmer waters fuel cyclonic activity. With the IPCC projection that storms can last longer from an agitated hydrological cycle, the chance for coastal flooding along well-eroded beach communities becomes accentuated. Taking factors like these into consideration, the EPA estimates that a 3-foot rise in sea level could enable the severe 100-year storms to occur on 15-year-storm frequencies—or at least more often than just once a century.

Additionally, a 1991 report by FEMA[198] estimated that a one-foot rise in sea level would increase the size of the 100-year floodplain in the U.S. from 19,500 to 23,000 square miles, and increase flood damages by 36-58%. If sea levels rise three feet, the 100-year flood plain would be increased to an astounding 27,000 square miles. Since coastal communities continue to grow each year with more and more people vying for real estate closer to a sandy beach, over time, there will be more people living along the coast, and subsequently in harm's way. Over the next century, the number of flood-prone households in the U.S. is expected to increase from approximately 2.7 million to possibly 6.8 million by the year 2100—that is, *if* current trends in population growth continue. If however 100-year storms become more frequent, then homes in flood-prone areas may loose their curb-appeal, thus slowing the rise in residents.

Knowing what we do now about the future of eroding coastlines, some states have enacted laws to prohibit new houses from being built in areas likely to be eroded within the next 30 to 60 years. Concerned about the need to protect property rights, Maine, South Carolina, and Texas have implemented various versions of "rolling easements," in which people are allowed to build along the coast, but only under the condition that they'll remove their homes when threatened by an advancing shoreline. Amazingly, people still pay top dollar for this kind of oceanfront property, although by the time the mortgage is paid off, the house may have to be removed completely.

Coastal marshes and swamps are also endangered from the rising seas as wetlands retreat inland, taking over once dry land that could have been used for construction, industry or farming. This also poses a threat of salinization of surrounding waterways and land, which can poison freshwater and turn farming fields fallow. New York, Philadelphia, and much of California's Central Valley get their water from portions of rivers that are slightly upstream from the point where water becomes salty during droughts. If saltwater from the rising seas is able to reach farther inland and subsequently upstream in the future, then freshwater in these regions will become tainted.

Life on the beach will change in coming years. This isn't a postulation for potential peril; it's a fact. Seas will continue to rise from thermal expansion alone. If glaciers continue to melt, and additional freshwater is released from enhanced precipitation of the hydrological cycle, sea levels will rise even more. There is little we can do to stop it. Yet we can prepare ourselves by building smart and knowing the floodplain dangers where we decide to live and build, while realizing that higher ground has a beauty all its own, and is not so threatened by the sea in a warming world.

SOMETHING'S FISHY

As a young boy growing up in the wooded Appalachia region of Pennsylvania, I learned at a young age the tricks of the trade to catching trout. These smaller cousins of the salmon occupy the streams, rivers and creeks in the land where I was raised. During a good part of the year, these intelligent, hard to catch fish provided more than just sport, they also were part of the diet in the region. Although I love eating the white, flaky, semi-sweet meat of these game fish, the sport of catching them was even more enjoyable. The thrill of hooking one of these sly swimmers kept me coming back to the water's edge to catch my daily limit time and again. Catching trout was unlike any other fishing in the area. These trout were intelligent creatures living in crystalline water where they could see approaching predators (including anglers) above them. They had a finicky diet as well, and didn't just nibble on any ole worm or lure. The fastidious nature of these Appalachian game fish made them difficult to capture, and more often than not, kept them at a safe distance from the dinner table. But their delicate nature also makes them less capable of adapting to a warming world.

While we often think of our oceans warming from climate change and sea levels rising along our coastlines, we may tend to overlook the waterways flowing in our own backyard. Just like all bodies of water around our world, streams, rivers and lakes will be touched by the warmth from our Sun, which will transform the world we've grown accustomed to—even the streams and creeks where I fished as a young boy. The biggest problem facing our local freshwater fish like trout, is that these species need a fluctuation in the seasons. Fish such as trout have evolved in conditions that become excessively cold in the winter, and somewhat warm in the summer. Warm up their waters, and the fish will do what they can to adapt. That's when things start to change.

As trout and other freshwater game fish become affected by warming waters, they will likely migrate northward in search of colder water. Doing so can result in disturbed breeding habitats, which can also lead to fewer surviving offspring, and in a worst-case scenario, extinction of some species. According to the EPA, an 8°F increase in annual air temperature, which is expected by 2100, is projected to eliminate more than 50% of the habitat of brook trout in the southern

Appalachian Mountains.[199] This will force the brook trout to move to a different place where they eventually could encroach on other waterways and the fish stock that live there. While this will further endanger these freshwater fish, the EPA does warn that cold-water fish habitats could be lost entirely in such states as Maine, Massachusetts, Connecticut, Ohio, and Nebraska, should global temperatures reach the heights anticipated by 2100. Presently, more than 750,000 people fish for trout in those states each year. One day though, their great grandchildren may only hear tall tales of "the one that got away" and may never have the opportunity to experience the thrill of the catch—unless they move to where the fish go, which is somewhat perplexing in a warming world.

As temperatures rise and fish head off to cooler waters, some fish will have a difficult time adapting to their new surroundings. In the Mississippi and Missouri River, fish can find colder water by swimming upstream. In the Great Lakes, fish can find colder water by diving to greater depths. Since the north to south flowing tributaries such as the Mississippi and Missouri rivers provide routes to northern and hence colder waters, fish within these large rivers have a chance to escape the heat. But for streams, creeks, and rivers flowing from east to west, fish won't be able to swim northward, and will be stuck in waters not favorable for their survival. Also, while some fish like salmon are able to migrate between inland rivers and the ocean and could easily move to more northern rivers, most freshwater fish can't tolerate saltwater and will never be able to make this journey on their own; thus, leaving many species of freshwater fish stranded.

There is something interesting though in putting together a scenario on diminishing fish populations in a warming world. Warm water promotes biological activity, which could result in the growth of more fish-food within the waterways. Warm waters could help vegetation and insects proliferate in a climate that sees less cold temperature. As mentioned back in chapter 2, mosquitoes like the pitcher-plant (*Wyeomyia smithii*), are now waiting longer to begin their winter dormancy, thus providing more juicy insects and larvae for fish to chow on. Nevertheless, scientists are still uncertain as to just how freshwater fish will react in these scenarios: the fish might leave because it's too warm, or perhaps they'll adapt with a more abundant food supply. Then again, the warm waters may attract warm-water fish like crappies and suckerfish, which will encroach on the coldwater fish like brook trout, forcing them to find another home.[200]

The amount of fish in this third scenario (warm-water fish replacing coldwater fish) would provide equilibrium to the overall population of fish, but would obviously have a detrimental impact on the coldwater fish. Moreover, I can speak from experience that I'd rather catch and eat a trout instead of a crappie or suckerfish. Yet even if we could adapt to catching and eating these warm-water fish, they too may become endangered from lower levels of oxygen in their new watery homes. According to one study, if we get up to the 8°F temperature that's predicted to occur by 2100, then the majority of southeastern rivers will have oxygen concentrations below the level necessary to support most fish.[201]

Not all fish around the globe would be adversely affected in a warming world though. Similar to the theory that warm water will enhance the bottom of the food chain in streams and rivers, the deep waters of the ocean may thrive as well. This would assist in the growth of deep-sea species, and possibly allow them to reproduce at younger ages, both of which would increase their populations. This could be countered though by a decline in deep ocean upwelling, where nutrients are brought to the surface to help feed the beginnings of the food chain from which all life in the ocean depends. The upwelling is a process similar to the THC, where currents are constantly moving around the depths of our ocean, and at times, surfacing from the deep on their long conveyor ride around the globe. Gravity tends to pull all of the good stuff to the ocean floor, yet upwelling brings these goodies to the surface where life begins anew. Since the THC would slow down though, many upwelling events would follow suit in a similar manner and decrease as well. This upwelling could also be affected by increased frequencies of El Niño events, which also disturb the natural rhythm of the ocean's currents and the bounty it brings forth.

Coastal marshes and the fish that call them home will also be affected by rising seas. Marshes, which are often thought of as swampy blights for breeding mosquitoes, provide a safe haven for newly born crab, shrimp, and other coastal kinds of seafood. These species may fare quite well as sea levels begin to rise, but over time their populations could be decimated. Crab, shrimp, and the other types of marsh-residing seafood species, reproduce in parts of marshy wetlands that are within 50 to 100 feet of the open water. These shallows provide them safety from fish that can't chance swimming into such low-lying waters. This and the abundance of food for the young crustaceans makes the shallow, swampy waters of coastal marshes an ideal nursery until the infant aquatic arthropods living there grow big enough to venture out into deeper waters. These marshes however sit on the tipping point of rising seas, being on the dividing line separating land and sea. As the ocean swells higher, these marshes, being only inches above sea level, will start to change, and so will the habitat for young crabs and shrimp.

As the seas rise, marshes will become inundated. At first, this will be in favor of the young crabs and shrimp since the increased sea levels will penetrate farther inland a bit to create more channels throughout the marsh, which increases the total area of marsh, thus providing an even bigger nursery for the tiny crustaceans born there. As sea levels continue to rise however, the marshes become deep enough that areas that were once swampy marsh become open bodies of water. This would then reduce the amount of coastal marsh that would provide protection for young crabs and shrimp, leaving them highly vulnerable to predators, which would inevitably lead to their decline, and in some cases, possible extinction.

Coastal marshes and wetlands are already declining, placing them at greater risk of extinction in the coming years as sea levels rise. Approximately 40% of the coastal wetlands of the lower 48 states are located in Louisiana, and this fragile environment is disappearing at an alarming rate. Louisiana has lost up to 40 square miles of marsh a year for several decades, accounting for 80% of the nation's annual coastal wetland loss. It's estimated that if the current rate of loss is not slowed, by the year 2040 an additional 800,000 acres of wetlands will disappear, and the Louisiana shoreline will advance inland by as much as 33 miles in some areas.[202] This loss is due to mostly erosion, but throw in a couple of hurricanes like Katrina and Rita, and the loss becomes even more profound.

Similar to the impacts of marshland, coastal estuaries will also be threatened by the rising seas. Oyster harvests will no doubt take a beating since parasites such as MSX and Dermo that require salty water, can more easily infiltrate estuaries that are overcome by intruding saltwater from the sea, thus reducing the freshwater content that would keep MSX and other such parasites at bay. As mentioned some chapters back when discussing the impacts of rising seas on the Chesapeake Bay, parasitic diseases such as MSX and Dermo have already taken a toll on Chesapeake Bay's oyster population, which have dwindled to nearly 1% of their historic levels. Many other estuaries have been seeing oyster decimation as well with declines in oyster populations as high as 90%.[203]

Oysters aren't the only seafood we'd need worry about from damaged estuaries though. Approximately 50% of all ocean fish, including flounder, sea trout, croker and red drum spend most of their lives in the ocean, but spawn in estuaries. As with the shrimp and crabs that find sanctuary for their young in marshland, these bigger fish rely on estuaries for their nurseries to keep their offspring safe from the mouths of predators. Damage the estuaries, and these fish will be placed in harm's way.

In some form or another, a warming world will transform the life within Earth's waters, having a tickle-down effect that will touch the lives of people around the planet. Food and recreation will be changed over time, yet the land around our world's waters will look different as well.

WOOD NOT

Plants love carbon dioxide; they thrive on the stuff. Without CO_2, trees, flowers, and all kinds of flora wouldn't flourish. One might think then, that increasing greenhouse gases like carbon dioxide into our planet's air would be a good thing, making our planet a greener place to live by providing Earth's vegetation with more of its much-needed CO_2 fertilizer. One might think of a warming world as a Garden of Eden—some kind of warm, balmy, lush tropical paradise where plants grow galore in a hotter world with an over abundance of carbon-based plant food. Unfortunately, this picture of a future world teeming with a bounty of botany is still under investigation with results varying a great deal from recent studies. If there's one thing we've learned from human history though, it's that whenever the hand of man disturbs the balance of nature, our zero-sum world can adversely compensate with detrimental side effects.

Skeptics have used the CO_2 fertilization stance as a fallback argument, often stating that *if* global warming is caused from human-induced greenhouse gases, then things won't be so bad. Senator Inhofe for instance, has used this argument in his Senate speeches, touting that adding more carbon dioxide into our atmosphere is doing our planet more good than harm. While plants do appreciate carbon dioxide to a certain degree, and perhaps that degree could indeed be very high, there does come a point when other factors in a warming world will counteract the benefits that can come out of a copiously rich carbon dioxide atmosphere. Recent studies and experiments known as Free Air CO_2 Enrichment (FACE) have shown this to be true.[204]

FACE studies pump various amounts of CO_2 around an area of trees or plant life and later, the trees and plants are analyzed to see how well they responded. These studies indicate that while foliage will respond initially to increased levels of CO_2, the productivity of the plants will be short-lived. This is primarily due to the fact that as CO_2 levels increase, other nutrients like nitrogen and phosphorous become more limited. Additionally, there may come a point of saturation where plants no longer respond to the increased levels of carbon dioxide. Yet other studies dispute this, and paint a much greener picture.[205]

Since most of the CO_2 fertilization studies have only been in research for a decade or two, it's far too early to conclude just exactly how increased levels of carbon dioxide will influence our world's trees, plants and flora on long term scales. We do know however, that warming climates and changes in precipitation in a heated world will alter the environments that certain types of trees and flora have evolved within.

As the Earth warms, it's believed that dry areas will become drier and wet areas will get wetter—drought-prone regions today will become even more arid, while elsewhere, rain becomes far more pronounced. In the drier regions that get less frequent rainfall—like regions in America's Southwest—woodlands that are now filled with trees may become desiccated tinder for forest fires as their limited water supply dries up. It is likely that many of these dry regions that see rain infrequently now will then become populated by grasses that require much less water to survive. This could change the landscape of many woodlands located in arid or semi-wet regions to eventually turn from forest to pastures. Fire though, is a concern of paramount proportions as we're feeling the impacts of this right now.

After seeing a steady decline in fires over a period of five decades, the northern regions of Canada have seen a marked increase in fire since about 1970. The area of northern forest burned annually in western North America has doubled in the past twenty years, which parallels the observed warming trend in this region.[206] Moreover, since the 1970s, firefighting and detection technologies have been greatly improved, yet fires are becoming more widespread and frequent. Similar trends have been noted in Europe and Asia as well, which points to the strong possibility that fires would be much more severe if it not for recent technological advancements, thus pointing to a correlation of increased global temperatures and forest fire ferocity. In a way, this is a blessing in a warming world; if the forests didn't burn and were left to flourish, then the forests would gather an over-abundance of fire fuel, which could then lead to catastrophic fires at some point in the future. Nevertheless, fire is more of a curse since it threatens homes, pollutes the air with ash, and can further decimate an already arid landscape.

A warmer and wetter climate though, would allow certain types of trees to flourish. Trees that are better at adapting to warmer conditions, such as oaks and pines, would prevail, and forests in regions supplied with sufficient or abundant rainfall could become denser. More notable to the landscape though will be mountain ridgelines that will climb to higher elevations, as warmer temperatures no longer prohibit tree growth from what was once an environment too cold for their survival. This will occur in higher latitude regions as well. This may leave many snowcapped mountains green as trees inch higher to the summit. Nevertheless, like just about everything else in climate change science, the effects on plant life in a warming world are riddled with uncertainties.

Climate models suggest[207] that, without including any beneficial effect from carbon dioxide, anywhere from 12-76% of the deciduous, leafy-tree forests in the United States would become thinner, while somewhere between 2-49% of these forests would experience increases by the year 2100. While these model simulations contain a wide, confusing, and inconclusive margin, the extremes point out that without accounting for any CO_2 fertilization, as little as 12% of our leafy tree forests would decline, and 2% would thrive, thus leaving a total minimum reduction of 10% in leafy tree forests. At the high end, the total reduction could be 27% and as high as 74%. Once again though, this doesn't mean our forests would turn brown; instead, other plant species that can thrive in warmer, drier conditions would likely take over.

To complicate the matter of forest impacts even further, when CO_2 fertilization *is* included in these simulations, a brighter picture emerges. Under initial carbon dioxide enrichment, less than 7% of these forests are expected to decline, and more than 92% are likely to increase. By adding in the factor of positive benefit from carbon dioxide, a radical difference is seen in these already perplexing models. Still, this benefit may be short-lived as demonstrated in the FACE research. While models conflict with the idea of flourishing forests in a carbon dioxide enriched atmosphere, it is important to note that not all plant species respond to CO_2 fertilization in the same way, which I'll discuss in more detail later in this chapter in *Feeding the Masses*.

As far as our forests go though, there is certainly a lot of uncertainty in predicting just what exactly will happen to our woodlands in the coming years, with no one just quite sure if our forests will grow, decline, or be replaced with other types of trees and flora. We do know however that our forests will change, but likely not disappear. Changes in the forests though will impact the wildlife that prosper in these woodlands. If wooded habitats shift to cooler areas in higher latitudes and higher elevations, many forms of wildlife could potentially adapt to global warming, just as they have adapted to the changes in climate that have occurred over millions of years throughout Earth's history. But humankind may stand in their way. In the past, when Earth was warmer and when it was colder, humans hadn't yet built roads, buildings, and other developments that blocked potential migratory routes.

Another concern of forests in a warming world is the infiltration of insects and arbor diseases. Climate models suggest that as our planet heats up, tree-eating insects will flourish, and become more widespread.[208] In recent times, we've already seen the damage that insects and parasites can have on our woodlands, and as time marches on, this will very likely worsen. From 1920-1995, there was a significant mortality rate for Northern Canadian forests with a loss of over 188 million acres of woodland, most of which occurred from 1970 to now as things have been heating up.[209] Similar trends have been observed in Russian forests where annual insect damage and disease mortality has affected nearly 10 million acres. The recent tree devastation from insects is so great that in Siberian and Canadian forests, insect damage is estimated to be of the same magnitude as fire loss. In earlier times, these regions were cold enough to keep most tree-eating insects, parasites and diseases from coming close to the woodlands of the north, but now that temperatures have increased, so has the range for these tiny tree killers.

A root-eating disease known as *Amillaria* is becoming more prevalent now as well since more regions can currently provide this disease with the 25°C temperatures it needs to survive. At present, Amillaria root disease is causing extensive loss in Canada's Pacific Northwest with a 43% decrease in annual volume of lodgepole pine, which is also under threat from bark beetles and other insects that can now make their way into higher latitude forests that are now warm enough to sustain their survival. Even if carbon dioxide fertilization were to help our trees grow, the infiltration and proliferation of insects, parasites and diseases could likely counter any positive effect.

The world's forests currently cover about 8.6 billion acres, or about 30% of our planet's total land area (excluding Greenland and Antarctica). About 57% of the world's forests are located in developing countries that depend on the trees for industrial wood products, fuel, and other non-wood forest products like mushrooms that grow under the shade of a forest's canopy, as well as nuts, fruits, palm hearts, herbs, spices, gums, aromatic plants, rattan and medicinal and cosmetic products. It's estimated that 80% of the population of the developing world—several million households—depends on just the non-wood forest products alone to meet some of their health and nutritional needs. If carbon dioxide can provide these forests with more fertilizer to stimulate and sustain plant growth for a long period, then populations in developing countries will fare well in a warming world. Given the odds against the trees in a warming world though, over time, these developing countries, already living in poverty, will take additional hits financially and nutritionally.

As the world heats up, forests will be transformed. We know of many likely scenarios, but uncertainty remains. Even in the best-case scenario where carbon dioxide fertilization promotes growth, other factors such as soil nutrient saturation, arbor-eating insects, parasites and fires could overshadow any benefits from the increasing amounts of CO_2. Research on trees in a warming world though is still ongoing, and in many respects, is just now beginning. Time will tell us more as we wait and watch for harbingers of climate change in the treetops.

HOME ON THE RANGE

The wide-open expanse of America's West is home to some of the Earth's most spacious, rolling rangelands. These meadows of grasses and shrubs became the residence of pioneers that settled in a home, "where the buffalo roam, and the deer and the antelope play…" These early American settlers, on their sojourn from the east, didn't find a land heavily forested by trees once passing the 100-degree meridian; instead, they found endless pastures of dense foliage, ideal for grazing sheep and cattle. Rangelands still comprise a vast portion of the American territory, accounting for about 40% of the landmass of the United States.[210] Today, a large portion of these rangelands is administered by the federal Bureau of Land Management (BLM), which leases grazing rights to private ranchers that raise a good portion of America's beef. The U.S. though is not the only country with these wide-open pastures; in fact, over 30% of the Earth's surface is range-land[211], which has a vulnerable side that at first might not seem obvious in a warming world: they have grass—and no trees—for a reason.

The grasses and shrubs that grow on the world's rangelands exist there instead of trees due to their low water requirement. Rangeland grasses require much less water for their survival, and are thus usually located in drought-affected regions, or regions that could endure drought-like conditions. In a warming world, drier regions may continue to get drier, which brings up a couple of important issues.

First, since rangeland grasses can withstand drought-like conditions better than trees and rich flora, they have a better chance of survival as global temperatures rise and rainfall becomes limited to many rangeland regions. Yet, since fire danger will also be on the rise in these arid regions, surviving grasses and shrubs would provide added fuel for large-scale grass fires, which could decimate great portions of the rangelands.

Second, if things get dry enough, then the grasses would of course wither, dry up and die, resulting in less pasture for grazing animals such as sheep and cattle. Surprisingly enough though, these two issues may become completely overshadowed by other deeds of humankind.

Rangelands face the threat of overgrazing by livestock. In the U.S., livestock actually outnumber humans.[212] While generous quantities of grains and feeds are used to fatten up our nation's cattle, a good deal of rangeland is used for grazing. Over the western eleven states, 73% of the publicly owned land, which is overseen by the BLM, is grazed. The combined acreage of public land that is grazed in sixteen western states is about 270 million acres, which is close to the total acreage of Oregon, Washington, California and Idaho combined. In all, the BLM manages about twice as much land as the U.S. Forest Service does. While this great expanse of land has been managed to a fair degree of sustainability in recent times, in the past, rangelands rapidly deteriorated as settlers, unaware of the damage they were imposing on the land, let their livestock overfeed in many areas; thus, one of the reasons why the BLM oversees much of America's rangelands today.

Before rangeland reform was strongly enforced in 1995 by the BLM, it was found that only 1/3 of the federally owned rangelands were in satisfactory condition, leaving an overwhelming 2/3 of these rangelands in only fair or poor condition. By the end of 1998, BLM assessments found that only 50% of these rangelands were in fair or poor condition—a slight improvement from the pre-1995 assessment. Yet data from the Society for Range Management indicate that about 15% of the BLM's federal rangeland is improving but about 14% is in declining condition as well.[213] While these rangelands take time to recuperate, wildfires that have been intensified from recent droughts continue to wipe out large areas of grasses and pastures. Additionally, other less desirable, opportunistic grasses such as broom snakeweed can infiltrate and choke out premium grasses that are grazed upon by sheep and cattle. While overgrazing can be kept in check to a certain degree with land management, other factors of humankind add stress to rangelands as well.

While the world population continues to grow, more and more rangeland comes under threat of annihilation from human development. In a 15-year period from 1982 to 1997, more than 12 million acres of privately held rangeland was shifted to other uses in the United States. During five years between 1992 to 1997, almost as much land was shifted into urban development as was converted to cropland.[214] While there is some concern how rangelands will react in a warming world from decreased rainfall in drought-prone areas as well as fire damage, the deeds of humankind hold just as much—if not more—of a threat to our wide-open prairies and pastures.

As rangelands lose their biologically diverse flora, their entire ecosystem changes. As grasses disappear, soil becomes vulnerable to erosion from both wind and rain. As soil disappears, it escalates into further rangeland loss as water storage from infrequent rains becomes diminished, which in turn reduces the chance for many grass species to regrow. This not only reduces the food supply for livestock; it also hinders the food supply of native animals as well. The sage grouse for instance, which was once abundant throughout Eastern Washington, from California to the Dakotas, and in the east and south to the Oklahoma panhandle, nearly made the endangered species list in 2004.[215] Conversion of rangelands into wheat fields and residential subdivisions, damage from livestock grazing, depleted water from big and thirsty cities like LA, Las Vegas, and Phoenix as well as increased fire frequency have all led to a dramatic decline in the sage grouse.

In many respects, the sage grouse is the canary-in-a-coalmine of the world's rangelands. Serving as a bellwether and barometer for the Earth's vast grassy fields and meadows, what's affected the sage grouse could eventually affect humankind as well. Although more resilient than forests in a warming world, rangelands have been under attack in recent times. Not only is a warming world imposing drought and subsequent fire to rangelands, the ravages of man in a world of escalating populations with ever more mouths to feed is stressing grazing land to a tipping point. As the planet heats up even more, humankind *and* a warming world could send our rangelands over the brink. Nevertheless, with land management underway now, we can hope that sustainability with proper planning will help maintain a balance of nature we can all live with. After all, our lives depend on it.

FEEDING THE MASSES

Since 1950, the world's population has more than doubled, going from about 2.5 billion people in 1950 to around 6.5 billion people in 2005. Today, there are more mouths to feed on our planet than at any other time in the world's 4.5 billion year history, and agricultural industries are doing a fairly good job at keeping up with the ever-increasing demand for sustenance. The world's population will continue to grow however, as will the need to feed a hungry planet. Most of the hearty staples that we rely on depend on climate though, and in a world faced with a *changing* climate, bringing food to the masses will require some alterations in the way we grow crops and raise livestock today. Some of these changes will not be as difficult for those who have the technology and resources, yet others who don't may be left behind.

In a warming world, the landscape of agriculture will shift. We've seen this already in rangelands where, if it not for land management, surely many pastures that are feeling the heat from recent droughts would be ravaged by overgrazing, and the subsequent loss of soil and nutrients from erosion. Droughts can further impact an arid region on the brink, bringing on wildfires that take dry range and farmlands into a flaming spiral of death. Yet we do live in a zero-sum world for the most part, and this brings some comfort, in that as one region faces the threat of desertification, devastation and decimation, someplace else will see increased rainfall and warmer temperatures to spawn more plant growth. As our planet transforms however, humans will need to adapt—that's the hard part.

Increasing temperatures in a warming world can have both positive and negative effects, especially on crop yields. Since various regions around the world will change differently as temperatures rise, some areas will see the positive benefits while others won't be so lucky. In the northern regions of the United States as well as Canada, warmer temperatures will likely increase crop yields, but crops in the already warm, low latitude regions of the southern U.S. will likely be hindered from the heat, resulting in a decline in crop yields there.

Another concern is how certain types of produce will react to the increasing temperatures. As with the FACE experiments with CO_2 fertilization mentioned earlier, there is similar evidence[216] that suggests a 2°C increase in temperature would help many crops to flourish and become more abundant. Thus, initially, with the CO_2 fertilization and small increases in temperature, many crops may do quite well. Yet further increases in temperature (like those predicted by 2100) have the same results as excessively increased levels of CO_2: plants hit a breaking point, and crop yields would then decline (if nothing were done to help manage the crops to withstand these climatic changes). Additionally, many of the experiments that have fared well in controlled laboratory conditions don't do so well in actual field conditions; thus, warmer temperatures and increased levels of carbon dioxide, overall, may do more harm than good—at least in some respects.

Crops such as wheat, rice and soybeans, known as C3 crops, seem to initially respond quite well to enriched CO_2 fertilization; however, C4 crops, which include maize, sorghum, sugarcane and millet don't respond as well.[217] In our zero-sum world, this would kind of equal out: more C3 plants than C4. But do we need (or should we have) more rice than sorghum, or more soybeans than maize? Additionally, how significant is the potential for C4 crops to become more vulnerable to increased competition from C3 type plants?

There's also a concern that longer summers may change timing for growing and harvesting seasons. And while some farms suffer the ravages of drought and put excess stress on dwindling water supplies, some regions will likely get heavily soaked by more rain resulting in runoff that could damage soils, and over saturate farmlands as well. There are many unknowns and a lot of *what ifs* when looking at the future of farming, but farmers do have a lot of options open to them, provided they have the financial and land resources available to allow nature to work in their favor.

Farmers can change planting and harvest dates, rotate crops, be careful on selecting crops that grow best in their newly transformed regions, and work towards better water management for their crops as well. At the same time, farmers could take steps to reduce greenhouse gases emitted from their yields, such as the methane emitted from rice patties and livestock, nitrous oxide from cultivated soils and feedlots, and carbon dioxide from the cultivation of virgin lands that once tilled, release carbon stores from the Earth. By planting trees to soak up carbon dioxide, and growing biofuels that could replace carbon dioxide-emitting fossil fuels, farmers could take further steps to thwart increased human-induced global warming, and thus help to manage not only their crops, but also the planet as a whole. Since cows alone produce about 80 million tons of methane annually around the world, finding ways to sequester any type of greenhouse gas from farming and ranching would be a step in the right direction—not that it would be easy to capture cow farts, but offsetting these emissions with reductions elsewhere in agriculture would certainly help.

All in all though, it doesn't look like the U.S. is in any danger of feeding itself now or at any time in the future—no matter how the climate will change. According to a report from the Pew Center on Global Climate Change[218], the impact of climate change on U.S. agriculture will result in some regional shifts and changes in the way we do farming, but nothing should stop us from feeding the hungry mouths in the land of the free and home of the brave.

Currently, nearly 400 million acres of farmland are cropped each year in the United States, and agricultural output has increased by about 2% each year since 1945. That means that today's agricultural output in the U.S. is nearly three times what it was near the end of World War II. While that's impressive in its own right, bear in mind that this increased output was accomplished as farmland declined. What's saved us and improved our food production, is technology. Since 1945, farmers have been using high-tech pesticides to protect plants from devastating critters, fertilizers have become more potent, and genetic engineering has made many plants more resilient as well. The United States has some of the world's best scientific minds constantly working on ways to better our food production. Most farmers also have enough capital to adapt to changes, shift their crops as needed, or invest in new technologies. But not everyone in the world is as lucky as we are.

Countries in Africa are growing in population while facing many problems including AIDS and drought. While fighting human diseases, many poor nations in Africa struggle with excessively arid conditions, which are decimating their farmland. Food production in most of Africa has not kept pace with the population increase over the past three decades.[219] In Africa, food consumption exceeded domestic agricultural production by 50% in the drought during the mid-1980s, and more than 30% in the mid-1990s. Food aid constitutes a major proportion of the food trade in Africa, and in many countries, it constitutes more than half of all net imports. In Kenya and Tanzania for instance, food aid constituted 2/3 of their food imports during the 1990s. Africa's current state is not at all self-sustainable. If aid were to be cut off to Africa, millions of people would die of starvation.

War has also played a role in Africa, displacing people from their homes and farms they could be tending. Combine these tribal feuds with ongoing, widespread diseases, droughts that have continued for decades and can be worsened as global temperatures rise, and you have a large portion of the world's population placed on the brink of catastrophe. According to the IPCC[220], major changes in African farming systems may compensate for some of the decreasing crop yields as climate change continues, but additional fertilizer, seed supplies, and irrigation will involve an extra cost—a cost that is hard to pay for in a land ravaged by war, drought, disease, unstable governments and ongoing famine.

In a warming world, it will undoubtedly become difficult to successfully feed the entire planet. We in the West can count our blessings for stable governments that provide land management, farm subsidies, technology for the advancement of agriculture, and the wealth of a nation that has advanced in its methods to feed not only its people but other nations as well. Yet we need to realize that not everyone will come out on top, and that some nations may face the slow and painful death of starvation. This could put additional strains on wealthy nations to provide increasingly more resources for underdeveloped countries lacking the resources for life, liberty, and the pursuit of happiness in a warming world.

HEALTH ISSUES

The Amazon is a beautiful place teeming with an abundance of life. Warm temperatures combined with more than adequate rainfall allow myriad plant species to thrive and prosper. The Amazon receives some of the highest amounts of precipitation anywhere on the planet: an amazing nine feet per year, which is a couple feet higher than the often-soggy Pacific Northwest's seven feet of annual rainfall.[221] This provides ample nutrition and shelter for more than half of the world's plants, animals, and insects that live there. The Amazon is so rich in biological diversity, that of the 3,000 plants around the world that are known to fight cancer, 70% of them come from the Amazonian jungles. In fact, 25% of the active ingredients in today's cancer-fighting drugs come from organisms found only in the rich forests of the Amazon.[222] Yet the beauty of the Amazon is deceiving as this lush Garden of Eden has a dark side that accompanies a warm and humid environment: infectious diseases.

Malaria, dengue fever, yellow fever, leishmaniasis and chagas disease are just a few of what are known as vector-borne diseases: diseases carried by mosquitoes, ticks or animals that can be passed on to humans. Many vector-borne diseases are prevalent in the Amazon, which provides an ideal breeding ground for these serious ills and the mosquitoes, ticks and other living things that carry them.

Normally, these tropical diseases remain in the latitudes that hug the Equator, but as the world warms, the survival habitat for these diseases will likely increase. It's expected that as temperatures rise, the range of vector-borne diseases like malaria and dengue will spread to higher latitude regions and to higher elevations in mountainous zones as well.[223] So although our planet will likely become greener as forests inch toward higher latitudes and elevations, diseases common in arbor abundant regions will be sure to follow close behind the northward advancement of flora.

Over the years, science and medicine have been able to fend off the spread and mortality of infectious vector-borne diseases like malaria. Yet even today, nearly 270 million people around the world are infected with malaria annually, and up to 2 million people die from the disease each year.[224] And as mentioned earlier back in chapter 2, dengue fever has spread in Mexico and Central America above its former elevation limit of around 3,300 feet to near 5,600 feet.[225] Additionally, the World Health Organization warns that many vector-borne diseases are spreading today due to our recently warmer climate and that at least 5 million cases of illness and 150,000 deaths each year are attributed to global warming alone.[226] This trend will likely continue.

It is true that in a zero-sum world, vector-borne diseases may drop off in some locations as those regions become drier. Yet the areas that will become drier have already been in drought-like conditions for many years or decades, and thus have never had much worry over such maladies as malaria or dengue to begin with. You rarely hear of a malaria or dengue outbreak across the great plains of the U.S.—it just simply hasn't been an environment conducive for these moisture-loving diseases. Planetary warming and increased precipitation will undoubtedly help the proliferation of disease-carrying pests though, and enough rainfall could result in a literal flood of diseases to some areas.

Since rainfall events are expected to last longer in a warming world, deluging the wet areas of our planet, flooding will likely become more frequent. Flooding and heavy precipitation not only provides large quantities of standing water where mosquitoes breed and larvae hatch, it can also disrupt sanitation systems. Floods are known to compromise sewers and waterways, resulting in contaminated drinking water. This then brings on other ills such as cholera and diarrhea. A typhoon that hit the Trust Territories of Micronesia in the Western Pacific in 1971 is a prime example of how flooding can result in catastrophic contamination. After the typhoon of 1971 hit Micronesia, freshwater supplies were disrupted, which forced people to use many different sources of groundwater. Unfortunately, the groundwater wasn't used very often, and over time, it became contaminated with pig feces.[227] This led to a widespread outbreak of an intestinal parasitic disease known as *balantidiasis*, which leads to severe diarrhea, cramping, and bloody stools. So while the people of Micronesia were wise enough to avoid the flood-contaminated above ground waterways, unbeknownst to them, their groundwater had become contaminated as well, leaving them with virtually no safe water to drink.

Flooding though is notorious for bringing on vector-borne epidemics, which was seen in Costa Rica's Atlantic region in 1991 after an earthquake created flooding and a malaria outbreak soon followed. The Dominican Republic in 2004 also suffered flooding, which similarly led to malaria outbreaks. Periodic flooding linked to El Niño has been associated with malaria epidemics in the dry coastal region of Northern Peru as well as the resurgence of dengue in the past ten years throughout the American continent. West Nile Fever has also resurged in Europe subsequent to heavy rains and flooding, with outbreaks in Romania in 1996-97, in the Czech Republic in 1997 and Italy in 1998.[228]

Flooding can also bring about other health risks such as drowning, injuries, and tetanus. Hypothermia in the wake of flooding is a concern as well, as are upper respiratory ills that can result from enduring such harsh conditions. And if floodwaters rise high enough, hydroelectric power can be disrupted, thus further hindering sanitation systems that rely on their electricity.

Heat waves will be another health concern in a warming world, such as those mentioned in chapter 1 where the heat wave of 1998 scorched Texas, claiming 100 lives there, and Dallas had to endure 100°F temps for 15 straight days. And more than 250 people died in the New York heat wave of 1999 that brought temps above 95°F for 11 consecutive days. And in recent times, the heat wave of 2003 caused thousands of deaths across Europe, with some estimates pointing to as many as 35,000 to 50,000 deaths from those abnormally hot summer temperatures that year.[229]

Florida and the Gulf Coast are also prime examples of dealing with the heat against the backdrop of global warming. In hurricane-devastated areas, ice is often found to be more valuable than water, since beating the heat is just as important as fending off dehydration. Once a hurricane rips through a region and takes out the power grid, the luxury of air conditioning is left as a mere memory. This leaves residents sweltering in tropical heat and humidity, and exposes much needed medications to damaging temperatures.

As temperatures rise, so will the occurrences of heat waves and the deaths they cause. According to the IPCC, by 2050, heat waves could result in up to several thousand extra heat-related deaths annually. This heat-related mortality increase would however be offset by fewer cold-related deaths in milder winters, which currently accounts for about 1,000 deaths each year in the United States.[230]

As with rangelands and agriculture that will likely change in a warming world, the effects on human health are issues that can be managed—provided the resources are available to do so. Prosperous nations will undoubtedly survive through adaptation with the continued advancement of science for treating diseases, proper land management to assist flood control, and improved emergency response for saving lives as well. Many things can be done to prepare for the health impacts of a warming world—but it takes money to provide the means to do so. Some people of this world will not have it so good, and will inevitably face dangers that threaten not only their way of life, but health and wellbeing as well.

THE INSURANCE ENIGMA

On the morning of January 17, 1994 at 4:30 AM, while most Southern Californians were still deep in slumber, the Earth no longer stood still. Twenty miles west of Los Angeles, deep below the city of Northridge, a fault gave way, causing a magnitude 6.7 earthquake that lasted only 15 seconds, yet killed 51 people, injured 9,000, collapsed nine parking garages, closed nine hospitals, left 25,000 homes uninhabitable, and collapsed nine freeway bridges.[231] All within mere seconds, with no warning—it just happened, in the blink of an eye. Within the aftermath of this tragedy, while in midstream of their obligations to assist those who entrusted them, the insurance industry rewrote the rules, and changed the way they do business in California. The insurance industry wrote a new chapter in our country's history of natural disaster recovery—a sad story that can affect everyone in our nation, especially in a warming world where veracious storms threaten regions more frequently, putting our homes and lives in harm's way.

It's estimated that the Northridge earthquake cost the insurance industry approximately $12.5 billion dollars.[232] The costs would have been much higher if everyone had a pricey earthquake protection policy at the time, since the total damage is estimated to have been between $20 billion to $44 billion.[233] One year after the quake, 93% of the insurance companies in California, after being flooded with claims and running out of money, severely restricted—or refused to write altogether—new homeowner policies. This triggered a crisis that by mid-1996 seriously threatened the vitality of the state's housing market and stalled the state's recovery from recession.[234] In attempts to avert bankruptcy in the insurance sector and to try and boost the curb-appeal of California real estate, in September 1996 the California State Legislature established the California Earthquake Authority (CEA), to provide homeowners with earthquake protection policies. This allowed the insurance companies to breathe a sigh of relief, and never again have to face giving homeowners any of their funds if the ground were to shake fiercely again—which inevitably it will. While insurance companies got off the hook from Northridge, the people of California have been left between a fault and a hard place—an unfortunate position that can befall us all in times of costly natural disasters.

Insurance companies in California no longer offer earthquake insurance, and the only game in town is the CEA. Either you have an earthquake policy through the CEA, or you play Russian roulette with the tremors of terra firma. The standard CEA policy has a 15% deductible. This means a homeowner must pay the first 15% of the value of his or her home, and in California, where a small home can easily cost $500,000, that comes out to a whomping $75,000 deductible before the CEA policy would kick in. These CEA policies, which have been more expensive than the policies previously available to Californians before the Northridge quake, are also limited to only $5,000 in coverage to personal belongings. If a homeowner loses all of their belongings in an earthquake, he or she will be given only $5,000 from the CEA to replace everything they own. And while the homeowner is waiting for their house to be rebuilt, he or she would get a paltry $1,500 from the CEA for temporary living expenses. If a family had to relocate to an apartment while their home undergoes repairs, they would be allotted only enough money to cover the cost of most one-bedroom apartments in Los Angeles for just one month. This amount might not even cover the move-in costs of an apartment rental either, especially in the Los Angeles or San Francisco Bay areas. If that isn't bad enough, the CEA policies also don't cover detached garages, pools, patios, landscaping, walkways or fencing. But that's just the tip of the insurance ignominy iceberg.

If "The Big One" shudders California, and a large portion of the state is hit so hard by a high magnitude earthquake that the CEA funds run out of money, the homeowners can be left up the proverbial creek without a paddle (or worthwhile policy). If such a widespread catastrophe were to happen, and numerous CEA insured homes were severely damaged, earthquake victims could receive pro-rated claim payments, perhaps as low as a dime for each dollar of covered damage—hardly enough to pay for expenses after doling out tens of thousands of dollars (or hundreds of thousands of dollars) in deductibles. In other words, if California's most common natural disaster were to happen again and if it reached a great enough magnitude, the residents of California would be left with practically nothing, resulting in extreme financial loss, which could in turn lead to a great exodus from the state, and a downward spiral for the West's economy.

While Californians pray the ground doesn't shake too hard, homeowners everywhere are faced with a similar dilemma. Insurance companies become profitable when they don't have to dole out their dough on damage. As long as the premiums keep coming in, it's like, well, money in the bank. Once the insurance companies have to live up to their end of the bargain however, all bets could be off—proven tragically all to well from the Northridge disaster, and the CEA bailout plan. In recent times, this is becoming a grave concern to everyone, and not just Californians.

Preliminary estimates of insured losses from Hurricane Katrina are around $34 billion, and total economic losses are likely to amount to more than $100 billion, making it the most costly and devastating natural disaster in the nation's history.[235] While many homeowners hit from Katrina have battled unsuccessfully with insurance companies[236], these policy providers will undoubtedly feel the financial repercussions from the heavily active 2004 and 2005 hurricane seasons. As waters warm, the threat of hurricanes may escalate. Even so, while it may be debatable whether hurricanes will continue to inflict damage as they did in recent years, storms in general will very likely increase in duration and ferocity, which is already hitting the insurance company as of late.

Europe insurers Swiss Re and Munich Re have calculated that 2005 was the most expensive year on record for natural catastrophes around the planet, with losses of more than $210 billion worldwide. Windstorm destruction in just the United States, the Caribbean and Mexico alone cost $83 billion in 2005, most of it coming from Hurricane Katrina. Additionally, the insurance giant AIG said in October 2005 that six of the ten most expensive hurricanes in U.S. history occurred in just the prior 13 months.[237] This should come as no surprise to the insurance industry, since this was foreseen some years ago.

Way back in 1991, long before the onslaught of Rita, Katrina and Wilma in 2005, the U.S. Federal Emergency Management Agency (FEMA) published a report[238] warning that as global warming kicks into gear, catastrophic damage would be highly likely, which in turn would severely impact the insurance industry. In their report, FEMA estimates that by 2100, insurance premiums for those living in hurricane-prone areas could increase by as much 200%. The biggest costs have been proven by hurricanes of recent times where wind and flood damage could easily bring yearly insurance costs into the tens of billions of dollars.

The IPCC estimates that, after accounting for inflation, annual global economic losses from catastrophic events increased from $4 billion in the 1950s to $40 billion per year in the 1990s. If you also include events of all sizes—not just catastrophic ones—the figure increases to about $80 billion.[239] Much of this could be attributed to the fact that more and more people are moving to coastal areas; thus, there are more structures placed in harm's way now than back in the 1950s. Nevertheless, insurance companies have and will continue to take a financial hit from storms in coming years—even if not related to climate change, more people will continue moving to hurricane-prone regions and continued development will place more property and lives at risk. Add in the fact though that it is very likely that our warming world will create storms with longer duration and heavier ferocity, and the future of the insurance industry looks a little bleak.

Insurance companies though will do what they have for years: take money from the pockets of people around the world in a ratio high enough to compensate for any losses they would have to pay out. Inevitably, insurance premiums to cover property will go up. Hopefully though, people in harm's way won't have to face the nightmarish scenario that California does, where if a big enough catastrophe should occur, the insurers merely slip through their loopholes and leave we the people with little hope for recovery.

Seeing the bellwethers of changing times and climate, we can hope for the best, yet prepare for the worst. The human species has done a great job at adaptation in a world that's been quite harsh throughout history. We can prepare coastal communities for upcoming rises in sea levels, plan the future of our fisheries and forests, manage our rangelands, adapt our agricultural practices for climate changes, advance medicines to stave off vector-borne diseases, and even the insurance industry could get smart and help the people they've promised to protect. It's a long road ahead, and much work needs to be done to prepare for each and every scenario. While preparation is paramount, prevention is just as important, and there are things we can do now before things really heat up.

8
Wean to Green

Nature will bear the closest inspection. She invites us to lay our eye, level with her smallest leaf, and take an insect view of its plain.
— *Henry David Thoreau*

For the sake of argument, let's assume that human-induced greenhouse gases are at least one contributor to our recent global warming trend. Seems like a fair assumption at this point. There's consensus in the scientific community, and simple physics tell us that the higher the concentration of greenhouse gases in our atmosphere, the hotter our world can become. We know that greenhouse gas emissions have increased significantly in recent times, especially since the Industrial Revolution, and we can feel confident that greenhouse gases are being pumped into our atmosphere at rates too high for nature to recycle. No matter what the outcome, whether we get more rain or less rain, more hurricanes or fewer hurricanes, more drought or less drought, I think it's safe to say that pumping our air with excessive amounts of carbon dioxide, methane, hydrofluorocarbons, perfluorocarbons and sulfur hexafluoride, probably isn't such a good thing. No matter what your stance on global warming, I think it would be safe to say that we all agree that it would behoove us all to think about reducing human-induced greenhouse gas emissions. After all, if nothing else, it couldn't hurt. Could it?

If we were all to stop pumping greenhouse gases into the atmosphere, our world would radically change. How would you get to work? Gas-burning cars would have to be sidelined. Could you work from home? Perhaps not, since the electricity fed to your house may likely be coming from a coal-burning power plant. And what would you eat? Food in your pantry came across highways and byways in the back of a diesel truck, which right now, is likely run on fossil fuels. Flipping the switch on greenhouse gas emissions would stop our world dead in its tracks. Instead of going cold-turkey, we need to wean ourselves off of greenhouse gas emitting fuels and other sources. This won't be an easy task, but if we address it now, we can build a better future. But who exactly should address this issue?

We know that about 30 billion tons of carbon dioxide are released into the atmosphere every year from humans alone. As mentioned in chapter 1, this is small in comparison to the total 600 billion tons emitted per year by humans *and* natural sources. Yet 30 billion tons is a lot of anything. If you figure that a good-sized pickup truck weighs about one ton, imagine stacking up 30,000,000,000 trucks into a great big heap. That's how much carbon dioxide humans are emitting into the atmosphere each year. Yet not everyone around the world is contributing to the increased concentrations of greenhouse gases in the atmosphere. It wouldn't be fair, for instance, to ask the Inuit to stop using snowmobiles while letting Hummers, Suburbans and Escalades speed along on America's freeways *en masse*.

So who exactly *is* to blame? Interestingly enough, it isn't necessarily the countries with the most people; instead, it's the countries with the most money. Topping the list of all prosperous, greenhouse-gas-pumping nations is the United States. The U.S. currently emits well over 6 billion tons of carbon dioxide into the atmosphere each year. The twenty-five states of the European Union come in next with combined emissions close to 4 billion tons. China is the second largest *single* nation pumping out greenhouse gases with nearly 4 billion tons a year. Russia, Japan, and India fall close behind.[240] Economies are fed, fostered, furnished and fattened by fossil fuels today, evident by the biggest emitters around the world like the United States, China, Europe and others. Cutting off the flow of greenhouse gas emissions right now would devastate world economies; thus, a slow yet efficient process must be pursued.

There are things that can be done to ease off the emissions of greenhouse gases, yet many are not that simple. Throwing in the towel and admitting defeat though is not an option. As humans, we've learned to adapt to changes in our world. But more importantly, we've learned to advance technologies to our advantage. By continuing our technological advancements with a focus on reducing greenhouse gas emissions, we can strive to change the future for the better, and continue to thrive in prosperous economies.

ALTERNATIVE ENERGY

Approximately 382,000,000,000 gallons of gasoline are burned each day in the United States.[241] At the same time, over 1,104,000,000 tons of coal are consumed each day as well[242], much of which is used to supply America with 50% of its electricity. The burning of fossil fuels like these around the world results in over 80% of all human-induced carbon dioxide emissions, or around 24,000,000,000 tons per year of CO_2. Gassing up our cars, heating our homes, and using electricity in any fashion requires a lot of energy and using fossil fuels to do it puts an overabundance of greenhouse gas into Earth's atmosphere. Still, our energy has to come from somewhere, and even though oil prices have increased in recent times (as they always have), fossil fuels are still the cheapest form of energy today. Oil and coal are abundant, and technology has been developed over decades to easily access it, transport it, process it and sell it. Switching over to something else now might not be so easy.

The vast majority of vehicles on the road today burn fossil fuels. It's not very likely that everyone will be willing to scrap his or her current vehicle for one that takes a more expensive fuel. Sure, it may be good for our environment, but who will be willing to make a costly sacrifice for a fuel that is not only more expensive than gas, but is also inconvenient to access as well? Cars are part of our lives; we often take them for granted. We depend on our vehicles to get us to work, the store, school, vacations, you name it—cars take us there. We don't have to put much thought into getting from point A to point B, we simply pull up to the nearest gas station, and within a couple of minutes we're filled with petrol and on our way. The ubiquity of gasoline stations combined with the relatively low price of gasoline, is one reason why we're now so dependant on this fossil fuel, and not likely to change to something else any time soon. Yet, changes are emerging.

Hydrogen fuel is seen as a viable energy alternative that could save the planet, yet there are no commercial vehicles as of yet that burn this simple (and highly flammable) gas. Hydrogen fuel though would significantly reduce greenhouse gas emissions since it releases practically none once it's burned. Just imagine—a guiltless fuel. Practically no carbon dioxide, methane or pollutants emitted at all, no matter how much we use. The trick however, is producing hydrogen fuel without involving fossil fuels in some form or another, which is still the least expensive method for manufacturing the clean-burning hydrogen fuel. There are other ways to make hydrogen fuel, but they do cost a little more than the fossil fuel methods right now. Over time, these processes will undoubtedly become comparatively less expensive as research and development continue and gasoline prices inexorably continue to rise.

The other downside to hydrogen is the Hindenburg effect—hydrogen is pretty darn flammable. While gasoline is quite combustible, hydrogen is extremely explosive, burning at a rate of eight to ten times faster than gasoline. Bear in mind that the H in H-Bomb stands for *hydrogen*. This stuff can be quite deadly if not handled properly. Thoughts of speeding down a freeway at 65 mph while carrying a mini Hindenburg-type bomb under your seat doesn't appeal to everyone. In many ways though, hydrogen is a good deal safer than gasoline or diesel, if (and this is kind of a big *if*) it is handled properly.

Because it is so light, hydrogen disperses and floats skyward when leaked—it won't soak into clothing, soil, or pool up on an open road after an accident like gasoline does. And because hydrogen is lighter than air, oil-spills like the infamous Exxon Valdez disaster would be a thing of the past since a hydrogen breech would result in the gas drifting skyward—provided it didn't ignite, which is the biggest safety concern with hydrogen. Yet a test conducted at the College of Engineering at Miami University, where 3000 cubic feet per minute of hydrogen was leaked from a vehicle tank and set ablaze, quells these fears by illustrating a unique and often unrealized safety in hydrogen. What researchers found in their vehicle-burning experiment was surprising.[243]

Over the course of the burn experiment at Miami University, temperature sensors inside the vehicle did not measure an increase of more than 2°C anywhere inside the vehicle, and the outside of the vehicle remained relatively cool as well, despite the fact that the leaking hydrogen fuel cell was flaming like a medieval dragon. The reason why the car remained cool during this experimental inferno is that when a carbon-based fuel like gasoline burns, glowing particles of hot soot transfer the heat to their surroundings—potentially including you. But because hydrogen contains no carbon, it burns cleanly without a residue of hot soot, producing little radiant energy. This means that a victim would have to be practically in the flame in order to get burned. This is also why all of the passengers on the Hindenburg who rode the burning dirigible to the ground, survived. As the Hindenburg became a flying inferno, the flames shot high into the air, allowing everyone in the passenger area—located on the underside of the blimp—to run for safety. Sadly though, 35 people jumped in panic from the burning craft before it landed, and died once hitting the ground. Nevertheless, the burning hydrogen didn't kill anyone riding the Hindenburg that day.

To set your mind at ease even more, the idea that you'd be driving a ticking H-bomb around just isn't so either. An H-bomb employs tritium—a fundamentally different form of hydrogen than that used in hydrogen cars—to replicate the same process by which the Sun generates energy in a nuclear kind of way. This occurs at astronomical temperatures and pressures where nuclear, rather than chemical reactions take place. Hydrogen fuels for cars are not contained at these kinds of pressures, nor do they use tritium.

In his 2003 State of the Union Address, President Bush announced a $1.2 billion Hydrogen Fuel Initiative to reverse America's growing dependence on foreign oil. Through partnerships with the private sector, the President's Hydrogen Fuel Initiative seeks to develop hydrogen technologies needed to make this alternative fuel practical, safe, cost-effective, and to be in place for use by 2020. The initiative will dramatically improve America's energy security by significantly reducing the need for imported oil, while at the same time taking giant leaps in the reduction of human-induced greenhouse gas emissions. Although we still need to wait and see if Bush's budget will continue to support the Hydrogen Fuel Initiative, without a start, there is no beginning to anything new. At least the wheels have been slowly spun into motion for a hydrogen fuel alternative.

It's hard to beat hydrogen in cutting back greenhouse gas emissions, but some other alternative fuels for transportation are in partial use today. Natural gas is becoming popular with about 130,000 vehicles in the United States today and over 5 million worldwide. Using natural gas instead of gasoline to power a car will cut back on carbon dioxide emissions by about 40%, but it will take a big bite out of carbon monoxide by cutting it by 90%, and nitrogen oxides are reduced by 60% as well.[244] In the past, using a car fueled on natural gas had a couple drawbacks. First, like all alternative fuel vehicles, the price was usually a bit more than your everyday gasoline-powered means of transportation. The second shortcoming was the availability of fuel—filling stations providing natural gas are rare. Both of these roadblocks though are being removed.

Americans can now get a tax credit on many alternative fueled vehicles such as the popular hybrids that use part gas, part battery for their fuel. Natural gas is now falling under that umbrella too, and in particular, car buyers opting for the natural gas-powered Honda Civic GX can get a tax credit of up to $3,600.[245] While this can soften the first blow of price, the second punch of availability is being overcome by a new invention from Honda called Phill.

Phill is a small unit that attaches to your garage wall and makes use of the natural gas that's already being pumped to your house—provided of course that you have natural gas at your home. Since so many residences already have natural gas lines in place—for use in heating, cooking, and hot water heaters—one doesn't have to go far to fill up their tank using Honda's Phill system. You just pull your car in the garage, hook up Phill, then settle in for the evening. You will have to wait a while for Phill to fill though, since natural gas lines in homes carry very low pressure, and cars like the Honda Civic GX require a pressure of around 3,600 pounds per square inch. It will take you a total of about six hours to top off one tank in your GX from Phill, but at least you don't have to plan your trips around a few stations in your county, or the hours they'd be open. It is important to note that Phill does use some electricity, and if that juice is coming from a coal-burning power plant, then there will be some greenhouse gas emissions in that part of the fuel chain. Still, Phill uses only 800 watts of electricity, and is about as quiet as running your run-of-the-mill clothes drier.[246]

An even more popular alternative to just the standard every day gas burning car is what's known as a "flexible fuel vehicle," or just FFV for short. You may be surprised to know that you might be driving one of these vehicles right now and not know it. It's estimated that there are over 5 million FFV vehicles driving around the American roadways today that can either use regular ole gasoline, or a combo-fuel of ethanol (grain alcohol) and gasoline. The popular blend of these two ingredients is called E85, which has 85% ethanol and 15% gasoline. Some of the cars sold today that can run on E85 include the new Chrysler Sebring, Dodge Caravan, Dodge Durango, Dodge Ram Pickup, Dodge Stratus, the Crown Victoria, the Ford F-150, Ford's Grand Marquis, the Lincoln Towncar, Ford Taurus, the Chevy Avalanche, Impala, Monte Carlo, Silverado, Suburban, Tahoe, GMC Sierra, GMC Yukon and Nissan Titan.[247]

Depending on the FFV you may drive, you could cut back carbon dioxide emissions by 30% or more.[248] Ethanol used in the E85 mixture is also produced domestically from crops like corn, sugar beets, sugar cane, barley and wheat, thus reducing the dependency on foreign petroleum sources. Yet, while a renewable, low greenhouse-gas-emitting fuel that reduces foreign oil dependency may sound attractive, it, like all renewable plant-based fuels, may not be feasible to provide a fuel substitute for an entire state or nation, and most definitely not the entire planet.

To make one gallon of ethanol you'd need about twenty-two pounds of corn. This means that you'd have to dedicate one acre of corn for about 328 gallons of ethanol. To provide the average American with enough ethanol for one year would require about eleven acres of farmland, which is also enough to feed seven people for a year. In a warming world where farms may be affected by excessive rain or drought, that's a tough trade-off. Should we fuel one car, or feed seven people? In all, if every automobile in the United States were fueled with 100% ethanol, about 97% of U.S. land would be needed to grow corn. If we used E85—using only 85% ethanol—we'd still need about 80% of all the land in the U.S. dedicated to the farming of corn for fuel (not food).[249]

Then there's the issue of processing ethanol. Planting, growing and harvesting just one acre of corn requires about 140 gallons of fossil fuels to run the tractors and other farm equipment used in the farming process. There's also additional energy used in the processing of the ethanol, which if using electricity from coal-burning power plants will further increase the amount of fossil fuel required to process ethanol.

No matter what we may rely on for renewable plant-based fuels for alternative transportation energy, we'll face the same problems as ethanol. Biodiesel faces these issues as well, since this fuel comes from such things as vegetable oil, discarded restaurant grease and animal fats. While biodiesel can provide a superb means of recycling waste products, there would likely not be enough used deep-fryer grease to go around, which would require farming of vegetables for oil extraction. Biodiesel does however provide some Earth-friendly features though, one of which is that it's biodegradable. If a boat running on biodiesel were to spring a fuel leak, the water would merely get a coating of vegetable oil, and not a toxic petroleum product—that is of course, if 100% biodiesel is used, which isn't always the case.

To run on 100% biodiesel, special engines or engine modifications are required. Standard diesel engines though can usually run on what's known as B20, which is a mix of 20% biodiesel and 80% regular diesel. In both cases though, carbon dioxide emissions are reduced. If using B20, carbon dioxide emissions are reduced by 15%, and if using pure, 100% biodiesel, carbon dioxide emissions are cut by 75%.[250]

While there are other alternatives to transportation fuel versus regular ole gasoline, many of these substitutes just aren't as feasible as hydrogen, ethanol or biodiesel. Propane still requires fossil fuel production, as its most abundant source is the byproduct from pumping out crude oil. Methanol has been considered in the past as well, but this form of alcohol is known to release high quantities of formaldehyde. Electric cars are a great idea, but since 50% of electricity in the U.S. today comes from the burning of coal, this isn't really a method of reducing carbon dioxide emissions—at least not yet. But if we were to address the fossil-fuel-from-electricity issue, then not only would the electric car become a more viable alternative, we'd also address a major contributor to carbon dioxide emissions from non-transportation fuel needs as well.

Making electricity from something other than fossil fuels is not an easy task. Hydroelectric power—coming from turbines spun from rushing water passing through dams—is a safe and effective alternative to coal, but you do need running water. In a warming world, we could use the expected higher precipitation levels in certain areas to our advantage. Currently, the U.S. only uses about 10% hydroelectric power to fulfill its electric energy needs, yet many untapped dams could possibly be used. Right now, only 3% of the 75,187 dams in the United States are set up to generate hydroelectricity—certainly, some of these dams could be retrofitted to provide hydropower. Worldwide, the idea of hydropower is catching on with 20% of all world power being provided by electricity-producing dams.[251]

Nuclear has been another popular alternative to coal-burning electric plants with 103 nuclear power plants in the United States today providing the country with 20% of its electricity.[252] As of January 2006, 30 countries worldwide were operating 443 nuclear power plants as well. Nuclear power is indeed clean with no greenhouse gas emissions. But there are many concerns based on the fact that we are dealing with similar technology that leveled Hiroshima to a barren wasteland. There have also been some disastrous calamities as well including the 1979 near-meltdown at the Three Mile Island nuclear power plant near Harrisburg, Pennsylvania, and the 1986 partial meltdown at the Chernobyl plant. And then there's the cost factor.

Today, nuclear power plants are not as economical to construct as coal-fired plants, largely because of the high cost of complying with regulations to restrict emissions of radioactivity. It's believed that if coal-burning power plants were regulated in the same way that nuclear power plants are, things would be quite different, and the costs would come down.[253] If you can bring the price down (or live with it), you would still need to do something with the nuclear waste, which could take anywhere from decades to a century to become stable enough to handle as regular refuse. Having increasing amounts of nuclear material building up over time also increases the terrorist factor—a threat none of us is willing to face, and would inevitably require nuclear power plants, storage and waste facilities to increase spending on security, which could also increase the costs of nuclear power.

There are many other small-scaled alternatives to energy including wind from giant windmills and solar power, both of which would be difficult to produce in large enough quantities to power an entire industrialized nation like the United States. Biomass has also been tossed around as an alternative, but this method has the same drawback as ethanol does in that it requires dedicated agricultural resources, which would be taken away from the hungry mouths on our planet.

While we look toward a future of reducing greenhouse gas emissions from fuels, there are many things that science, research, and the willingness to change can do to make our world a little cooler. There is however more than one way to skin a cat—instead of concentrating entirely on what fuel we burn, we could capture the gases from the fuels we burn today so that they can't get into the atmosphere in the first place.

SEQUESTERING THE OFFENDERS

Ah, Paris, France—a lovely, romantic city, home to the Eiffel Tower, The Louvre, the cathedral of Notre Dame de Paris, and an overwhelming abundance of culture, gourmet food and exquisite wine. The *city of light* is one of the most beautiful cities on our planet, yet back in the Middle Ages, this place was a filthy, disease-ridden cesspool. Sewage, including human waste, was carried along open gutters in the streets to the Seine River, which, ironically, was also the city's primary supply of drinking water. Chamber pots were emptied directly into streets, and new courtesies evolved where gentlemen, when escorting ladies, positioned themselves closest to the street, thereby placing themselves (rather than the ladies) nearer to where the sewage would hit the ground after being thrown out of second-story windows.

The odor in Paris became unbearable in many places throughout the city, and the unsanitary conditions were indirectly responsible for the plague, which wiped out a huge percentage of the population in 1348 and 1349. For hundreds of years, Parisians dealt with a nasty, human-induced byproduct (human waste and sewage) that just wouldn't go away. They had no way to contain it. A lifestyle had evolved around the workings of the city, and the residents of Paris couldn't just change an entire thriving society overnight. Nevertheless, awareness and perseverance searching for answers to the problem paid off, and in the mid to late 1800s, reliable sewers were built, and the city lived happily ever after—so to speak.

Parisians became aware of a human-induced problem, and identified the source of their ills as a byproduct of their lifestyles. In the case of Paris in the Middle Ages, the human-induced byproduct was feces, urine, and other wastes. Technology was developed to subdue the problem—not by eliminating the waste (that wouldn't be so easy), but by capturing and processing the byproduct, thereby eliminating the adverse effects of the byproduct into their environment. In many ways, this is no different from the issues we face today with human-induced global warming. We have a human-induced byproduct that if not captured, will have an adverse effect on the environment. So, instead of just letting greenhouse gases like carbon dioxide wisp through the air all willy-nilly, why not capture these gases and process them? It's a good idea, and one that is under way right now, although like anything else that takes technological change, it has some hurdles to overcome.

The process of capturing carbon dioxide is called *carbon capture and storage*, or just CCS for short. This process is primarily targeted at large power plants where the carbon dioxide waste is usually absorbed in a liquid and then stored someplace safe. Oftentimes, the CO_2 is pumped into abandoned oil wells and other underground sources, or is pumped into the deep ocean. Both methods have their pros and cons, with the biggest concern being that over time, we may be burying large quantities of carbon dioxide, which will inevitably—at some time or another—have to come out and face the music. Since it will take decades though for automakers, fuel suppliers, and electric companies to switch over to alternative non-greenhouse gas emitting sources of energy, carbon capturing could be used as a temporary means to reduce carbon dioxide emissions into the atmosphere now and allow us to play catch-up in technology. This would be a short holdover, kind of like cesspools in Paris would have alleviated the exposure of disease and nauseous fumes until a permanent solution (e.g. sewers) could be put in place.

CCS technology is already in use today, but it has a long way to go to make a noticeable dent in carbon dioxide emissions around the globe. Since 1996, the Norwegian company Statoil has been stripping about a million tons of CO_2 a year from the Sleipner West oilfield under the North Sea and injecting it at high pressure into water-absorbing rock formations known as *saline aquifers*. An even larger project that began in 2000 takes CO_2 from North Dakota and sinks it into an old oilfield in Weyburn, Saskatchewan, Canada.[254] These two instances prove that CCS is viable, but it comes at a price.

Besides the equipment needed to capture, transport, and store the carbon dioxide emissions, the capture and compression process to sequester it requires a lot of energy, so much so that power plants would need to increase their energy requirements by as much as 40% to implement carbon capturing.[255] This in turn would increase the price of electricity to the plants' customers, which might make the CCS process less appealing to the public.

While the case of the Statoil carbon dioxide capturing uses underground resources, there aren't enough of these big holes in the Earth to go around, which leaves the next possible CO_2 resting place in our oceans. This could be done in two ways. The first is by pumping carbon dioxide into the ocean at depths of around 3000 feet, and let the carbon dioxide just merely dissolve. The second method, called *lake-type* is similar, but the depth is greater (9000 feet or so), which would allow the carbon dioxide, which is heavier than water, to form a lake of CO_2 on the ocean floor, which would delay the inevitable dissolution that takes place in the first method. But since we'd be pumping massive quantities of carbon into Earth's life-giving oceans—a world of myriad ecosystems—this may not be such a good thing.

Large concentrations of carbon dioxide can kill ocean organisms in many ways, one of which is by increasing the acidity of the seawater. It's argued that the dark abyssal depths of the ocean contain barely any life at all, but further studies will need to be conducted to ensure we don't do more harm than good. In any case, many environmentalists are concerned that CCS won't be used as just a temporary fix, and that coal-burning power plants may find this to be cheaper than switching to alternative energy sources like nuclear power or hydroelectricity for long-term solutions.

Even if we did capture copious quantities of carbon dioxide and pump it into the oceans or deep underground, it still won't solve all our problems. The CCS technology is only geared towards power plants—not cars. Only about a third of the world's CO_2 emissions come from the generation of electricity. But, if we do capture some of the carbon dioxide, strive to reduce emissions, and develop technologies for alternative fuel sources, this triple whammy would no doubt lead toward a future of much lower greenhouse gas emissions.

Besides just capturing carbon dioxide, we could also work to recycle it through carbon sinks like trees and the ocean. Trees love carbon dioxide, and through photosynthesis, absorb and store the carbon from CO_2 and release healthy, life-giving oxygen. Trees though do this best when they are young, so new forests would need to be planted to help in the reduction cause. As good as this may sound, it would still take a lot of trees to do the job. In fact, to reduce carbon dioxide in the United States by 7%, it would require the planting of new trees over an area the size of Texas every 30 years.[256] This becomes even more complicated when you figure that if we were to use renewable energy like ethanol that requires colossal acreage for farming, if we were to grow enough corn for fuel, as well as trees for carbon recycling, we'd nearly run out of room to live. Still, every little bit helps, and planting some new trees sure couldn't hurt—provided large amounts of fossil fuels weren't used in the planting process to defeat the purpose.

Oceans are also great carbon sinks. Those coccolithophores mentioned previously, as well as all types of phytoplankton just love devouring carbon dioxide. These organisms can be stimulated by iron particles such as hematite, which come in microscopic sized particles. Natural sources of ocean iron have been on the decline in recent decades, which may be attributed in part by changing ocean currents and the lack of upwelling events that would normally bring iron-rich nutrients up from the ocean floor. By fertilizing the ocean with iron, we may be able to stimulate plankton production, which in turn would result in a greater sink to soak up carbon dioxide. This was tested in 2002 in the Southern Ocean around Antarctica, which showed that anywhere from 10,000 to 100,000 carbon atoms were eliminated for each iron atom added to the water.[257] Still, anytime we fool with Mother Nature, we tend to get burned. Skeptics argue that this iron enrichment may have adverse effects on plankton blooms, and much more research should be conducted before pursuing this recycling alternative.

Carbon dioxide, while being the biggest offender on the greenhouse gas hit list, is just one gas that could be captured. Other gases however could not only be captured, but also used to our advantage.

WASTE NOT, WARM NOT

Parisians in the Middle Ages may not have realized it, but they were emitting high concentrations of greenhouse gases into the atmosphere. Oh, don't get me wrong; it wasn't anything like today's emissions, but they were emitting greenhouse gases nonetheless. Although they lacked SUVs and coal-burning power plants back in the 1300s, the excessive abundance of excrement floating along in the open-air street gutters in Paris released the same offensive gas that termites and cows produce: methane. While carbon dioxide can be captured and contained, it's a bit more difficult to sequester the fumes from grazing cattle, wood-eating termites, or even we humans. Yet there are steps that can be taken to not only reduce emissions from some greenhouse gas emitting points of origin, but to use some of these emissions to our benefit as well.

As you may recall from chapter 1, about 34% of all human-related methane emissions come from our landfills. Landfills slowly decompose our trash, and in the process of rotting, methane is emitted into the air. While the gas emitted from landfills is about 50% methane, the other remaining 50% is almost entirely carbon dioxide. If these gases could be captured, similar to the CCS process, we'd make great strides in reducing a large portion of greenhouse gas emissions. While capturing the CO_2 from landfills faces the same issues of storage that the CCS approach has from power plants, methane capturing can provide a fuel source.

Unlike carbon dioxide, methane is a fuel and the primary component of natural gas, which burns much cleaner than fossil fuels. And as mentioned in the previous section, natural gas reduces carbon dioxide emissions by about 40% compared to the use of gasoline, which makes methane (or natural gas) an attractive alternative fuel. By letting landfills just sit and rot, millions of tons of natural gas are released into the air each year—wasted like the proverbial spilled milk. If we can capture that methane gas though, not only would we reduce greenhouse gas emissions, we could also use it to replace the burning of fossil fuels in some applications and lessen our dependency on foreign oil at the same time. Fortunately, as we speak, methane capturing and processing is underway.

The U.S. EPA is one outfit that's taken action on landfill gas reuse with a program called the Landfill Methane Outreach Program (LMOP). This program, which was established in 1994, encourages the recovery and use of landfill gas as an energy resource. The LMOP forms partnerships with communities, landfill owners, utilities, power marketers, states, project developers, native tribes, and non-profit organizations to overcome barriers in developing landfill gas processing facilities by helping to assess project feasibility, find financing, and market the benefits of project development to the community. As of December 2005, the LMOP has gained nearly 500 partners that have signed voluntary agreements with the EPA to develop cost-effective landfill gas energy products, and nearly 400 operational landfill gas projects are underway in the United States right now as well.[258]

The LMOP has had positive effects, and in 2005, the EPA estimates that the landfill gas energy projects in the United States prevented the release of approximately 20 million tons of greenhouse gases into the atmosphere. This reduction has the same equivalent environmental benefit as preventing the use of 162 million barrels of oil or offsetting the use of 341,000 railcars of coal—all from burning methane released from landfills.[259] Greenhouse gas concentrations are still on the rise, but they would be worse had it not been for the LMOP partners' actions to process landfill gas into usable energy.

There are numerous success stories among the EPA's LMOP partners, and to list them all would require a whole other book dedicated to the hundreds of projects in use today. A few outstanding successes though[260] include the Alachua County-Gainesville Regional Utilities company of Florida, which generates 2.4 megawatts of electricity from methane extracted from the nearby Southwest Alachua Landfill. These 2.4 megawatts provide enough electricity to power about 1,500 homes, all from the extraction of landfill-produced methane that would have normally just wisped away into thin air and accumulated in the already rich concentrations of greenhouse gases in our atmosphere.

The Antioch Community High School in Antioch, Illinois is another great example of landfill gas turned to useful energy. This school, working in conjunction with the local Edison Company, saves approximately $100,000 in electricity by partially heating the school with micro-turbines that are fueled by methane from the nearby H.O.D. landfill. A similar heating process is used by Cargill, Inc. of Fayetteville, North Carolina, where steam boilers are heated by burning methane from the nearby Cumberland County Landfill. Cargill then is able to process soybean oil from these methane-heated boilers, which can then be used as biodiesel, plastics, building materials, and food products.

Middlesex County, New Jersey though has taken a big bite out of landfill gas emissions by extracting enough methane from three of its landfills to provide an amazing 20 megawatts of electricity, which is used to power its wastewater treatment plant and portions of its electric grid. The amount of electricity produced by burning methane from these three landfills would be enough to power 12,700 homes. To look at it another way, the methane that's captured and processed by Middlesex County would be equivalent to planting 30,500 acres of forest, removing the emissions of 21,400 vehicles, or preventing the use of 260,000 barrels of oil.

Besides methane, there are other solid waste products that are linked to greenhouse gas emissions. Interestingly enough, much of this comes from simple things like paper that, if recycled, can significantly reduce the amount of carbon dioxide and other greenhouse gases emitted into the air each year.

Making goods from recycled materials typically requires less energy than making goods from virgin materials. And when energy demands decrease, fewer fossil fuels are burned and less carbon dioxide is emitted into the atmosphere. The EPA estimates that increasing our national recycling rate from its current level of 27% to 35% would reduce greenhouse gas emissions by over 12 millions tons versus just throwing this stuff away into a landfill. Waste prevention also makes an important difference: by cutting the amount of waste we generate by just 5%, we could reduce greenhouse gas emissions by at least another 10 million tons. Together, these levels of recycling and waste prevention slash emissions by more than 20 million tons—an amount equal to the average annual emissions from the electricity used in 12 million households.[261]

As with the advances in technology that made Paris a better smelling, cleaner city, research and development of practical applications for alternative fuels and greenhouse gas recycling (or capture) are steps that fine-tune our world towards a stable environment that we all can enjoy for countless generations to come. But while we pursue alternative fuels, carbon capturing, and various methods of recycling, we can't forget that emissions themselves need to be cut, and as our world population grows, this becomes harder to do, especially when our most basic needs and economies produce greenhouse gases today.

THE KYOTO PROTOCOL

So we know that greenhouse gas emissions are on the rise and we know of some ways to slow them down and eventually eliminate them. But who is going to make sure this happens? After all, not everyone is that concerned about a warming world, and if it means giving up the family Suburban for a tiny Honda Civic GX or Toyota Prius, people will likely be even more reluctant to change. But what if the governments of the world got together, came up with a plan for reducing greenhouse gas emissions, and vowed to keep a promise to do so? It may work, especially if it meant helping industries develop alternative fuels and methods to sequester or recycle the emissions, while not adversely impacting the way of life we've grown accustomed to—at least not too harshly anyway. An agreement like this could mean a cooler world in the future, but at what cost would this come?

In December of 1997, an international UN treaty was negotiated with the goals of reducing greenhouse gas emissions around the world. This treaty, known as the Kyoto Protocol, since its inception, has been more controversial than global warming itself. The idea behind the Kyoto Protocol is that developed, industrialized nations responsible for the increase in greenhouse gases since the onset of the Industrial Revolution need to drastically cut their greenhouse gas emissions, while developing countries that weren't producing so much of these emissions would stabilize their greenhouse gases. In theory, and on the surface, this might sound like a great idea—and it is; at least with good intentions at the heart of it all. But the devil's in the details, which makes this such a controversial matter.

There have been heated political "my team" debates going on over the Kyoto Protocol in the United States for years, ever since President George W. Bush said he would not ratify the agreement, yet only a few years prior, the Clinton Administration signed Kyoto. Bush's actions started a war of words on not only Capitol Hill but in bars and office water-coolers around the nation. People were either praising Bush for sticking to his holstered Texas six-shooters, or they were calling him an Earth hating conservative. Well, once again, we face what seems like a binary situation where either team A is right or team B is correct; however, there are many other considerations to this complicated issue that reveals answer C: neither side in this debate is entirely right.

Signing the Kyoto Protocol is nothing more than symbolic. Not until a country ratifies it does this treaty become legally binding. On November 12, 1998, the Clinton Administration did sign Kyoto. This was only a symbolic gesture though, performed by then VP Al Gore and Senator Joseph Lieberman.[262] The Clinton Administration never took the crucial step of submitting the protocol to the Senate for ratification however, and for good reason.

On July 25, 1997, before the Kyoto Protocol was to be negotiated, the U.S. Senate unanimously passed by a 95–0 vote the Byrd-Hagel Resolution (S. Res. 98). This resolution states that the United States should not be a signatory to any protocol that did not include binding targets and timetables for developing as well as industrialized nations or "would result in serious harm to the economy of the United States." Additionally, economic analyses, prepared by the Congressional Budget Office and the Department of Energy (DOE), the Energy

Information Administration (EIA), and others, demonstrated a potentially large decline in America's GDP from implementing the Kyoto Protocol. President Clinton then approved and signed into law appropriations bills for the years 1999, 2000 and 2001 that prohibited the EPA from using its funds to:

> ...issue rules, regulations, decrees, or orders for the purpose of implementation, or in preparation for implementation, of the Kyoto Protocol.

Clinton clearly could see that the Kyoto Protocol was a good idea, but it would have a devastating impact on the United States' economy; thus, he never pursued ratification of the treaty, and instead, set in motion policies to ensure America's economic safety from treaties like the Kyoto Protocol. While many blame Bush for being anti-environment and far different from his predecessor Bill Clinton, both presidents do seem to agree that ratification of the Kyoto Protocol, in the way it's currently drafted, would not be in the best interest of the United States, nor would it be fair.

One of the driving economic factors that has been a decision maker against ratifying Kyoto, is that under this treaty, only industrialized nations like the U.S. would have to reduce their emissions, and developing countries only need to stabilize theirs. While this at first may sound logical, bear in mind that China and India, according to the Kyoto Protocol, are not industrialized nations, and being considered developing countries, are exempt from the emission reduction requirements of the treaty.

China and India, according to Kyoto, can continue on with business as usual since it's felt that, even though they contribute largely to the world's total greenhouse gas emissions today, neither country did so during *most* of our last century—only recently. As such, China and India, according to Kyoto, are not viewed as being responsible for the build-up of greenhouse gas emissions that are causing global warming now. Still, today, the U.S. emits around 6 billion tons of carbon dioxide, yet China is the second largest single-nation emitter pumping nearly 4 billion tons of carbon dioxide a year into the atmosphere; India meanwhile emits a little over 1 billion tons per year.[263] The United States though, according to Kyoto, would have to reduce its total greenhouse gas emissions by 5% below 1990 levels, and accomplish this reduction by 2012. Seeing as though the protocol went into effect in 2005, to achieve this goal would require reductions today well in excess of 10% or more with no new emissions above what we produce now.[264] This kind of dramatic reduction, although a great idea for our planet, would undoubtedly devastate the U.S. economy while China and India would keep growing with no reductions in greenhouse gas emissions at all.

President Bush has followed suit with the position of the Senate, the Clinton Administration and the various economic reports indicating that the U.S. would take a financial hit while other countries like China and India can practically ignore Kyoto. Bush, stating his stance on Kyoto said:

> This is a challenge that requires a 100 percent effort; ours, and the rest of the world's. The world's second-largest emitter of greenhouse gases is China. Yet, China was entirely exempted from the requirements of the Kyoto Protocol. India and Germany are among the top emitters. Yet, India was also exempt from Kyoto…America's unwillingness to embrace a flawed treaty should not be read by our friends and allies as any abdication of responsibility. To the contrary, my administration is committed to a leadership role on the issue of climate change…Our approach must be consistent with the long-term goal of stabilizing greenhouse gas concentrations in the atmosphere.[265]

At the same time, India and China maintain that the major responsibility of curbing emissions rests with the developed countries (like the U.S.), which have accumulated emissions over a long period.[266] Yet, neither India nor China need worry about negative fiscal impacts from Kyoto, and can continue growing their economies. The United States however, wouldn't fare so well.

If the U.S. were to abide by the Kyoto Protocol, some studies show that U.S. productivity following implementation of Kyoto would fall by $100 billion to over $400 billion in 2010, and that gasoline prices would increase by 30-50% while electricity jumped up by at least 50% in cost. American workers would suffer as well with reductions in wage growth of 5-10% a year, while living standards would fall by 15%.[267]

India and China though would continue to produce energy at much cheaper rates than the U.S., which would give them the upper hand in many industrial sectors since their operating and manufacturing costs would be much lower than the United States. In a day and age of outsourcing, where millions of jobs are lost in the U.S. and Europe and are going overseas to China and India, not only would jobs continue to flood to foreign lands under Kyoto, but surely the economies of India and China would rise, as would their greenhouse gas emissions.

Australia has also refused to sign or ratify the Kyoto Protocol for similar reasons as the U.S., stating that it would cost the country jobs and hurt its economy. Australia, interestingly enough, isn't even among the biggest offenders, coming in at only the 11th largest emitter per capita of greenhouse gases. Still, many nations have come on board with a total of 160 countries as of January 2006 adopting the Kyoto Protocol, yet some of these signatories are clouded in controversy.

Russia, which ratified the treaty in 2004, is on board with the whole greenhouse gas reduction thing, and has no beefs about it either. But greenhouse gas emissions tend to rise at the same rate as economic wealth and prosperity. Since 1990, the economies of most countries in the former Soviet Union have collapsed, and their greenhouse gas emissions followed suit. With little or no money for gasoline or coal, there just isn't as much fossil fuel being burned nowadays around Russia. As such, its current emission levels are substantially below the targets set by the Kyoto Protocol. Not only does this set the bar quite low for Russia, it also gives them a unique economic advantage.

Kyoto allows for something called *emissions trading*. This is where, when one country meets its reduction requirements, it can sell off the remaining reductions to someone else. Russia for instance, according to Kyoto, has a 0% reduction rate—it just can't produce any more emissions above what they were in 1990. Let's say then that Russia finds their emissions are 5% below this zero-rate target. They then could sell this surplus 5% goal to someone like the U.S., who may not be able to meet their goals and emitted too much greenhouse gas during the same time. This emissions trading was set up to reward countries for meeting or beating their emission reduction targets while at the same time providing financial incentives to others to do so as soon as possible. But when a country is already practically in a deficit of emissions like Russia, then it is much too easy for them to sell their surplus of percentage points, while making it harder for countries that are truly trying to reduce emissions for the good of the planet (and not financial gain).

There's a similar point of controversy in the European Union. Although the EU has ratified Kyoto, emission levels of former Warsaw Pact countries who are now members of the EU have already reduced their emissions from economic restructuring, just like the former Soviet Union. Thus, the total emissions from all of the EU countries is evened out by economic disparity between rich countries like Germany and England, and by the poorer former Warsaw Pact countries like Poland, Romania and Czechoslovakia.

Then there's the issue of how much of a reduction is fair. While the European Union is required under Kyoto to reduce their emissions by 8% from 1990 levels and the U.S. has to cut back by 7%, Russia, as mentioned earlier, has a 0% reduction requirement, but surprisingly, some countries can *increase* their emissions. Iceland is one such country that, under Kyoto, can pump an extra 10% more greenhouse gas into the atmosphere from its 1990 levels. Norway can also increase its emissions by 1%.

The Kyoto Protocol may have its problems, but it's no doubt a step in the right direction. Kyoto sets an example, which has been adopted by non-Kyoto pacts around the world. The Asia Pacific Partnership on Clean Development and Climate is one such Kyoto-like agreement that has been signed by Australia, China, India, Japan, South Korea, and the United States. It was introduced in 2005 and allows participating countries to set their own goals for reducing greenhouse gas emissions individually, but with no enforcement mechanism. Supporters of the pact see it as complementing the Kyoto Protocol while being more flexible, yet critics have said the pact will be ineffective without any enforcement measures.

Additionally, many local governments are voluntarily adopting Kyoto-like standards, including nine Northeastern U.S. states involved in the Regional Greenhouse Gas Initiative (RGGI), which is a state-level emissions capping and trading program. It is believed that the state-level program will indirectly apply pressure on the U.S. federal government by demonstrating that reductions can be achieved without being a signatory of the Kyoto Protocol. Participating states include Maine, Massachusetts, New Hampshire, Vermont, Rhode Island, Connecticut, New York, New Jersey and Delaware.

As the Kyoto Protocol falls victim to political ramblings and heated debates, the push for lower emissions continues and a trend has begun. One day, we may see more laws or policies that require reductions in greenhouse gas emissions. Yet at the same time, we all have to ask ourselves a difficult question, "How much am I willing to sacrifice to impede the onslaught of a warming world?"

On a Personal Level

On January 20, 1961, a speech was made that would resonate in the hearts, minds and souls of Americans and many people around the world for decades, if not for centuries or millennia into the future. Some of that speech goes as follows:

> The world is very different now. For man holds in his mortal hands the power to abolish all forms of human poverty and all forms of human life. And yet the same revolutionary beliefs for which our forebears fought are still at issue around the globe—the belief that the rights of man come not from the generosity of the state, but from the hand of God...

United, there is little we cannot do in a host of cooperative ventures. Divided, there is little we can do—for we dare not meet a powerful challenge at odds and split asunder...
Let both sides seek to invoke the wonders of science instead of its terrors. Together let us explore the stars, conquer the deserts, eradicate disease, tap the ocean depths, and encourage the arts and commerce....
All this will not be finished in the first 100 days. Nor will it be finished in the first 1,000 days, nor in the life of this Administration, nor even perhaps in our lifetime on this planet. But let us begin....
And so, my fellow Americans: ask not what your country can do for you—ask what you can do for your country.

John F. Kennedy, the 35th President of the United States, spoke those words during his inaugural address. Reading them today, some 45 years later stirs my soul and brings a tear to my eye. Kennedy uttered those inspiring words as he looked towards a future of the United States, where we the people would do the undoable, achieve greatness and strength, and never give up hope. But what made this speech all that more moving, was the fact that Kennedy pointed out that each and every person could make a difference if we all row in the same direction and work together. Making a difference in our world, a world that's warming, takes this same kind of dedication, teamwork and perseverance.

While we as individuals may not be able to sequester carbon dioxide or build plants that run on landfill methane or ratify treaties like the Kyoto Protocol, there are things each and every one of us can do to help reduce greenhouse gas emissions.[268] While these things may appear insignificant, remember that every little bit helps, and collectively, seemingly irrelevant steps can result in monumental leaps and bounds.

The first steps you can take start at home, in particular, cutting back on energy use. This doesn't mean you have to turn your thermostat down barely above freezing or go without air conditioning; instead, you can choose to purchase products displaying the ENERGY STAR® label. Most ENERGY STAR labels require that products exceed minimum federal standards for energy consumption by 13-40%, depending on the particular appliance. Look for the label on refrigerators, washing machines, dishwashers heating and cooling equipment, televisions, computers, and home entertainment equipment. A high-efficiency refrigerator can reduce carbon dioxide emissions by 450 pounds a year. An energy-efficient washing machine can reduce emissions by 440 pounds a year. While 450 pounds of carbon dioxide may seem like a drop in the bucket compared to total emissions per year worldwide, if just half a million homes reduced their emissions by 450 pounds—from using an energy efficient refrigerator—the total reduction of carbon dioxide would come out to 112,500 tons, just from using a more energy-efficient refrigerator.

Insulation is another great way to save energy without noticing any change in temperature in your home or hot water. And if you live in a sunny climate, you might consider installing a solar thermal system to help provide your hot water, which could reduce your home's carbon dioxide emissions by 720 pounds a year.

Planting trees is an Earth-friendly gesture that not only beautifies your home, but also provides a great carbon sink. Deciduous, leafy trees are the best bet as they work as not only a great carbon sink—taking in about 50 pounds of carbon per year per tree—but also provide shade to keep your home cool in the summer, thus reducing the strain and power from using the air conditioner. And while maintaining your yard, if a push mower is out of the question, (which would cut carbon dioxide emissions by 80 pounds per year) consider using a composting lawnmower, which reduces the waste stream sent to your local landfill.

Recycling, as mentioned earlier, cuts back on greenhouse emissions as well since making goods from recycled materials typically requires less energy than making goods from virgin materials. By recycling aluminum cans, glass bottles, plastic, cardboard, and newspapers, you can reduce your home's carbon dioxide emissions by around 850 pounds per year. Also, you can promote recycling efforts by buying food and other products packaged in reusable or recyclable packaging. This simple action could reduce your emissions by 230 pounds a year.

Of course, using our automobiles wisely can significantly reduce greenhouse gas emissions as well. Simple, timesaving trips that combine your errands into fewer jaunts back and forth to the store can save gas and emissions, as will occasionally walking or riding a bike. When you do drive, keep your car tuned up and its tires properly inflated—both will save on fuel and hence emissions. Carpooling is another alternative, which if done just twice a week can reduce your carbon dioxide emissions by 1,590 pounds per year (on average). If your boss allows it, consider telecommuting or working flex-hours that allow you to drive during off hours from traffic. And if purchasing a new vehicle, consider a fuel-smart car—one that gets more miles to the gallon than your current vehicle. The potential carbon dioxide reduction for a car that gets 32 miles per gallon is 5,600 pounds per year.

By taking small steps, we can all work towards reducing greenhouse gas emissions. Our individual contributions to the cause may seem insignificant, but bear in mind that you'll be part of a bigger system where alternative fuels are being developed and implemented, greenhouse gases are being captured and sequestered, landfill gases are being recycled, and treaties are being drafted and ratified to ensure a safer environment for the future of humankind.

9

Mythical Questions

o o o o o o o o o o o o o o o o o o o o o o o o o o o o o o o o o o o o o o o

Myths and creeds are heroic struggles to comprehend the truth in the world.
—*Ansel Adams*

Shortly after writing the articles on global warming that inspired this book, I received a flood of emails with questions regarding the subject of climate change. I understood all too well the profusion of confusion surrounding the subject of global warming, and felt that by providing this book, I could help to sort out fact from fiction, rumors from truth, and science from superstition. I know however, that when it comes to finding a quick answer, it's not always convenient to flip through a volume of pages to find pieces to a puzzling problem. While I've attempted to address the questions I've received from my readers throughout this book, as well as other important issues surrounding climate change, I would like to dedicate this chapter as a kind of recap to lightly touch on each of the more commonly asked questions sent to me concerning global warming.

Mythical Question #1: Don't volcanoes emit more carbon dioxide than humans?
Answer: This is the most frequently asked question I've received from my readers. The truth of the matter though, is that this is really in large part just an urban legend. I discussed this briefly in chapter 4 in *Erupting Hazards* when talking about solar dimming, but I'd like to cover it again for a moment here as well.

Volcanoes do indeed release massive magnitudes of matter into our atmosphere, some of which includes greenhouse gases. But, volcanic eruptions account for only about 0.16 billion tons of carbon dioxide annually, which is diminutive compared to the 30 billion tons that humans pump into the air each year.[269] In fact, volcanoes tend to have just the opposite effect from carbon dioxide emissions. Instead of adding to global warming, volcanoes attribute to global dimming—a cooling type of effect.

The matter spewed forth from volcanoes is extremely high in sulfur, which combines with water vapor in the stratosphere that forms sun-blocking clouds of sulfuric acid droplets. These clouds then block sunlight from hitting the Earth. As you may recall from chapter 4, this was found to be true after the eruption of the Mexican volcano El Chichon in 1982. While this was a much smaller eruption than Mount St. Helens in 1980, the Sun dimming effect created by El Chichon's eruption was much greater due to the profuse amount of sulfur gases that it released. El Chichon pumped out far more sulfur gas than Mount St. Helens did, which resulted in El Chichon lowering global temperatures by three to five times as much as Mount St. Helens' eruption. Mount St. Helens' enormous eruption resulted in a dimming that cooled the Earth by a mere 0.1°C, while El Chichon's smaller eruption dropped temperatures by as much as 0.5°C in some areas.

It's also important to note that matter released by volcanoes needs to be shot up into the atmosphere high enough so that it doesn't get washed away when it rains. In most cases, the ash, dust, sulfur and other debris released by volcanoes do not make it past the troposphere where rain clouds form.

Mythical Question #2: Nature itself releases immense quantities of greenhouse gases into the atmosphere. Aren't the additional greenhouse gases added by humans insignificant in comparison?

Answer: It is true that nature releases a lot of carbon dioxide into the atmosphere. In fact, as mentioned in chapter 1 in *When Termites Fart*, humans release about 30 billion tons of carbon dioxide into the atmosphere each year, which is only 5% of the total 600 billion tons released from all sources. Mother Nature is responsible for 95% of emissions. But a couple things need to be considered here.

First, carbon dioxide remains in the atmosphere for a very long time, anywhere from 50 to 200 years in some cases. So every year, as we keep pumping an extra 5% of CO_2 into the atmosphere, much of it just keeps accumulating. Currently, we have a CO_2 concentration of 372 ppm in the atmosphere compared with only 280 ppm before the onset of the Industrial Revolution. In other words, year after year, as we've pumped more carbon dioxide into the atmosphere, a lot of it kept building up. Mother Nature's sinks (like trees, oceans, etc.) do a good job at recycling the usual 570 billion tons of naturally released carbon dioxide, but our extra emissions have accumulated for a total increase of nearly 33% today. According to the IPCC projections, that level will likely increase to 540 to 970 ppm by the year 2100 for an increase of 90 to 250% above what we had in the 18th century.

The second issue regarding our contributions of greenhouse gases compared to that in nature concerns the dramatic increase we've added in such a short period. According to the ice core sample data from Vostok mentioned back in chapter 5, we only came close to our current 372 ppm CO_2 concentrations once before around 323,000 years ago when carbon dioxide concentrations got up to 300 ppm—making today's concentration of carbon dioxide a record breaker. Additionally, when carbon dioxide levels did increase, they weren't as likely to increase at the rapid rate that's occurred since the Industrial Revolution to now.

Mythical Question #3: Carbon dioxide is good for plants, so what's wrong with adding more of it to our atmosphere?
Answer: I discussed this a bit back in chapter 7, and it's a very good question. Skeptics have used the idea of CO_2 fertilization as a kind of fallback argument. It's usually used as a safety net in skepticism by stating that *if* global warming is caused from human-induced greenhouse gases, then things won't be so bad. Senator Inhofe used this argument in his Senate speeches as well, peddling the idea that adding more carbon dioxide into our atmosphere would do our planet good.

While plants do utilize carbon dioxide to a certain degree, there does come a point when other factors in a warming world will counteract the benefits that can come out of an excessive carbon dioxide-rich atmosphere. The FACE experiments mentioned back in chapter 7 indicate that while foliage may respond initially to increased levels of CO_2, the productivity of the plants is short-lived, and other nutrients like nitrogen and phosphorous become more limited.

It is important to note however, that some studies dispute the FACE research regarding forest growth, but we do know that not all plants respond so well to CO_2 fertilization. Crops such as wheat, rice and soybeans, known as C3 crops, seem to initially respond quite well to enriched CO_2 fertilization; however, C4 crops, which include maize, sorghum, sugarcane and millet don't react as well. And while initially, some plants may respond well to CO_2 fertilization, as concentrations rise, there comes a tipping point where all crop yields could decline.

Then there's the issue of how temperatures and water supplies will inevitably change as global warming continues. This will transform many regions that today get limited rainfall. These regions will likely become even drier, and thus not have the most vital element for plant growth: water.

And lastly, a warming world—brought on by the same carbon dioxide that's argued as being good for plants—will face the issue of insect and plant disease proliferation. As our planet heats up, insects and plant diseases will be able to migrate to what is now a cold climate, which will threaten myriad forests that have never before seen these insects or parasites. We're seeing this now in fact, with details of the devastation laid out in chapter 7's *Wood Not* section.

Even if carbon dioxide fertilization were to help our trees grow, the threat of fire, changing temperatures, shift in water supplies and the proliferation of insects, parasites and diseases could counter any positive effect.

Mythical Question #4: Since the Earth's rotation around the Sun varies, isn't the Sun responsible for global warming?

Answer: A lot of this was covered in the section *Once Around the Sun* back in chapter 4, but it is a very important issue, and an excellent question as well.

While it is true that Milankovitch cycles vary our distance from the Sun and the tilt of our planet varies as well, these cycles are quite mathematically predictable. Taking into account the elliptical cycles of Earth around the Sun, we're now in a period of relatively small solar variations—nothing dramatic enough to explain recent temperature increases. Since the Younger Dryas ended some 11,500 years ago, we've been, for the most part, in what's known as an interglacial period, where things tend to warm up a bit, but not to the degree we have recently. When working the math on the Milankovitch cycles and our current epoch, we don't find any evidence pointing to a Sun-cycle trend that would be causing our anomalous warming today.

Another point to consider regarding links to Sun cycles and global warming, is that there is a direct correlation to increases in carbon dioxide since the Industrial Revolution to the rise in temperatures since then. Since Sun cycles take thousands of years, and our rise in temperature has only occurred in less than two centuries, the correlation to carbon dioxide increases seems more plausible, especially given the fact that we're not in a Sun cycle that could produce the results we've seen since the Industrial Revolution.

Mythical Question #5: Wasn't the Medieval Warm period warmer than today?

Answer: I briefed over this in chapter 2, and I'll elaborate more on that here.

In short, the answer is no. The year 2005 was the hottest year on record, with temperatures likely not this warm for over 10,000 years. The global average surface temperature came in at 58.3°F in 2005; that's 1.4°F warmer than it was 100 years ago. The Medieval Warm period was also not worldwide, and affected mainly areas neighboring the North Atlantic. While those areas saw a quite noticeable warming period, the Northern Hemisphere mean temperature estimates conducted by three separate peer-reviewed studies show temperatures from the Medieval Warm Period to be about 0.2°C warmer than those from the Little Ice Age.[270] In contrast, during the last century alone, we've seen temperatures increase by 0.6°C.

These temperature readings though are taken from the debated hockey stick graphs mentioned throughout the book. But as mentioned in *Hockey on Thin Ice* in chapter 4, it is interesting that nearly all graphs, whether made by skeptics or proponents of the hockey stick, show a warming trend since the onset of the Industrial Revolution.

Additionally, recent reports also show that the warming trend we're in the midst of is the longest recorded in the past 1,200 years.

Mythical Question #6: According to NOAA reports, hurricanes are only recently becoming intense because of natural cycles. Isn't it true that decadal oscillations are to blame for the increase in hurricane activity and not global warming?

Answer: This is one question that unfortunately doesn't have a simple yes or no answer. I discussed this quite a bit in chapters 5 and 6, and I'll cover some of the highlights here.

NOAA did publish an article[271] in November of 2005 where they concluded that the recent upswing in hurricane activity was "not related to increases in greenhouse gas warming." Their explanation was that a decadal oscillation known as "the tropical multi-decadal signal" is to blame. According to NOAA, a positive cycle of this oscillation began in 1995. This is signaled by the various tropical patterns I mentioned back in chapter 5 as well. NOAA's methods for hurricane prediction based on these factors are a well-accepted, accurate method of seasonal forecasting. But what they didn't mention in this particular article was an index they use in the forecasting process, which shows something anomalous about the recent hurricane trend.

Using a strength index called the Accumulated Cyclone Energy index (ACE) NOAA can observe decadal oscillations related to hurricane seasons. The ACE shows a busy, positive cycle from 1950 to 1970, then a slowdown into a negative cycle from 1971 to 1994. And it certainly shows an active, positive decadal cycle starting in 1995 to now. NOAA mentions that the positive cycle from 1950 to 1970 started to repeat itself in 1995 after it flip-flopped to a negative cycle from 1971 to 1994. This makes perfect sense. This is what multi-decadal oscillations do; however, the ACE index shows that our recent positive cycle that started in 1995 is stronger than the cycle from 1950 to 1970.[272] So while we indeed appear to be in a positive decadal cycle, this cycle is stronger than those we've previously seen in the past.

This corresponds to the research I mentioned in chapter 5 conducted by Kerry Emanuel of MIT and Webster's team of researchers from the National Center for Atmospheric Research and the Georgia Institute of Technology. Both of these published, peer-reviewed studies indicate two things.

First, there is a link to recent increases in sea surface temperatures and hurricane intensity, and we know that sea surface temperatures are on the rise from global warming. Emanuel's study used a similar index as NOAA's ACE, which found that using data since the 1930s, the total Atlantic hurricane power dissipation has more than doubled in the past 30 years, and in the Western Pacific, tropical cyclones have increased their power dissipation by 75%. Additionally, Emanuel reports that the duration of storms in both the North Atlantic and Western Pacific has increased by around 60%, and that the wind speeds of storms increased by about 50% as well.

The report by Webster's team had similar results showing a large increase in the number and proportion of hurricanes that reach categories 4 and 5. It is important to note that Webster's research didn't find any global trend in the *number* of tropical storms; instead, the *strength* of these hurricanes has escalated. In fact, according to Webster's research, the number of category 4 and 5 hurricanes has almost doubled around the world.

The second important point coming from these recently published studies is that if a multi-decadal oscillation were the only cause, then we would expect to see a decrease in activity elsewhere. Our planet tends to be a zero-sum world where anomalous conditions in one place are seen as just the opposite someplace else. Any upswing in hurricane activity say, in the Atlantic should be reciprocated by weaker activity in another part of the world. The increases found in these studies however are worldwide, and not just focused on a single point on the planet.

Another thing to consider in all of this is that global sea surface temperatures are on the rise, which is attributed to human-induced global warming. This is hurricane fuel. It may be just one element of the formula, but it is an important one nonetheless. While decadal oscillations will no doubt affect our world's weather and subsequently cyclone activity, warmer than normal sea surface temperatures caused by our warming world are most certainly an intensifier of hurricanes.

Mythical Question #7: In the beginning of the 20th century, a large meteor smashed into Siberia. Now, studies show this may have caused our global warming. Could this be?
Answer: A press release in 2006 from the University of Leicester in the U.K. broadcast this theory to the media.[273] Many news organizations didn't carry this story since its theory is riddled with holes. The news outfits that did pick up on this story though did so with sensationalistic gusto.

This study, which is not officially published in a scientific journal as of yet (or peer-reviewed as far as I know) hypothesizes that an intense meteor explosion known as the Tunguska event could be responsible for global warming during the

20th century. The Tunguska event was huge, and is estimated to have been around 10 to 15 megatons, which felled over 60 million trees over an area of 830 square miles in Siberia. This was a massive meteor impact, and at first glance, it would stand to reason that anything this intense would have some kind of effect on our planet.

The theory that was stated in this press release is based on the idea that water vapor was disrupted in the atmosphere when the meteor came crashing to Earth. Water vapor—like ordinary clouds—is indeed the best heat-trapping greenhouse gas around. If the meteor could perturb the atmosphere enough, it is possible that water vapor could be accelerated. But this doesn't really, well, hold water.

Since this incident happened in 1905, one would expect that the implications of this kind of impact would be felt immediately. After all, when a volcano erupts, we feel the solar dimming effects right away. But according to climatic records[274], nothing happened within even a few years of the Tunguska event. And here we are, over 100 years later. To think that a water vapor disruption would take a century to finally take hold just doesn't make sense. Even the effects from volcanic eruptions last only months to a couple years at most.

And there are other problems with this meteor theory as well, which are covered a bit in the next mythical question.

Mythical Question #8: Throughout the 20th century there have been numerous nuclear tests above and below ground. Isn't some of this linked to global warming?
Answer: This falls under much of what was discussed from the previous question, but has been used as an argument to support the Tunguska event theory as well.

First, as with the Tunguska event, we'd expect to see some kind of immediate impact to the Earth's climate. Also, as far as a heat trapping gas is concerned, nuclear fallout gets trapped in the troposphere and eventually comes down in rain, just like volcanic material. The shockwave though is another story.

One theory[275] has it that nuclear tests have sent shockwaves into the stratosphere that vented water vapor from the lower troposphere where clouds form. Thus, according to this theory, the high altitude stratosphere became filled with Sun-blocking water vapor, which actually *cooled* the planet—instead of heating it. The scientist behind the Tunguska event theory uses this to his defense, stating that this is why it took so long for the global warming to occur from the giant meteorite: nuclear tests cooled our planet for decades, but when nuclear tests stopped, the water vapor went away and we heated back up.

Nuclear tests would then be seen as more of an Earth cooling agent, and not something to heat it up. But given the fact that nothing happened immediately following nuclear explosions or the Tunguska event, and that we do have solid correlations between carbon dioxide concentrations and temperature increases, it stands to reason that neither nuclear testing nor the Tunguska event have a practical scientific basis on explaining global warming.

Mythical Question #9: Why won't President Bush support the Kyoto Protocol and address global warming?
Answer: There is a common misconception surrounding Kyoto's acceptance in the Presidency. In chapter 8, I dedicated a section entirely to the Kyoto Protocol to discuss its good and bad. To summarize some of the highlights, neither President Clinton, the U.S. Senate nor President Bush ever ratified the Kyoto Protocol. While it was *symbolically*, signed by Al Gore and Senator Joseph Lieberman, the Clinton Administration never submitted the protocol to the Senate for ratification.

Bush didn't ratify the Kyoto Protocol, yet neither did Bill Clinton. In fact, Clinton approved and signed into law appropriations bills that prohibited the EPA from using its funds to support Kyoto. Bush, Clinton, and the Senate have all seen that Kyoto, the way it is drafted now, would be unfair, and cause severe economic impact to the U.S. while countries like China and India wouldn't have to cutback on their emissions at all (the exact details on this are in chapter 8's section *The Kyoto Protocol*).

Mythical Question #10: There have been recent reports that the global conveyor and the Gulf Stream are slowing down, which could lead to an ice age. Is this true?
Answer: A recent study[276] published in the journal *Nature* in 2005 by Professor Harry Bryden from the School of Ocean and Earth Science University of Southampton in the U.K., was heavily broadcast throughout the media, stating that the Gulf Stream was slowing down. This would seem to conflict with other findings mentioned in chapter 3's *Salt, Heat and the Great Conveyor* that discuss a possible speeding up of the thermohaline circulation (THC) of which, the Gulf Stream is a small yet important contributor. This study however, has some oddities about it.

First, the level of uncertainty is high in this study, which brings into question a reliable link to tie these findings to global warming. In fact, the uncertainty is nearly high enough to cancel out the actual measurements that were taken. Secondly, if the Gulf Stream had slowed by 30% as presented in this study, then we'd likely see a great cooling across Europe, which hasn't occurred yet.

The most notable issue with this study though, is that while the researchers found a 30% slowing in the southbound currents, they found no slowing in the northbound currents. This could fall under the issue of uncertainty tolerance, but it is suspicious nonetheless.

We do know that more fresh water has been pumped into the Atlantic (and other oceans around the world) in recent times from meltwater, and that the salinity content of the Arctic region has become diluted as a result. This would lead to a slowing of the THC, and as mentioned in chapter 3, could lead to ice-age-like conditions eventually. Yet, other studies show that salinity levels were lowest in the mid-1990s, and that the seas have become saltier since then.[277] Nevertheless, as a whole, the oceans have become less salty in recent decades.

To confuse the matter even more, some studies, while agreeing that some northern seas might be less salty now, show that northward currents are compensating for this by flowing saltier waters to the north.[278] The increased salinity coming from the south though could be a harbinger of global warming, since in a warming world, more evaporation would occur in the Tropics, resulting in saltier water there. If so, then perhaps there is something strange going on with the THC, and that perhaps the southerly flowing cold water is less salty from melting ice, and the northward current is saltier from increased evaporation from the Tropics. Could this mean that as indicated in chapter 3 that the THC could be temporarily speeding up, which could then lead to a slowdown? Or could this also mean that the northbound THC currents are speeding up and the southbound currents are slowing down, both of which could be harbingers of an accelerated shutdown of the THC?

At this point, it's hard to tell. We do know that the THC (at least the northbound currents) appeared to be quite positive around the 2004 and 2005 hurricane seasons, which could contribute to hurricane formation from its warm water transport. We also know that current projections from the best climate models don't seem to think we'll see a complete shutdown of the THC in at least the next 100 years, although it will likely start to slow down throughout this century. Bryden's study could be on track with this projection, but further monitoring of the THC will be necessary to see how this critical component of Mother Nature's Goldilocks Complex will affect our climate in an increasingly warmer world.

To be continued...

www.GreenhouseTruth.com

Notes

[1] Planet Ark (Dec. 5, 2005) "Arctic Peoples Seek UN Help to Slow Warming"
Retrieved Dec. 7, 2005 from:
http://www.planetark.com/dailynewsstory.cfm/newsid/33808/story.htm
Reuters Alertnet (Dec. 5, 2005) "Pacific islanders move to escape global warming"
Retrieved Jan. 28, 2006 from:
http://www.alertnet.org/thenews/newsdesk/L05770367.htm
Mercury News (Dec. 3, 2005) "Inuits transformed by global warming"
Retrieved Dec. 6, 2005 from:
http://www.mercurynews.com/mld/mercurynews/news/breaking_news/13319240.htm

[2] Wikipedia, Scientific Skepticism
Retrieved 7/5/2006 from:
http://en.wikipedia.org/wiki/Scientific_skepticism

[3] BBC New Online (Nov. 22, 2000) "'No acceleration' in Pacific sea rise"
Retrieved 1/24/2006 from:
http://news.bbc.co.uk/1/low/sci/tech/1035489.stm
Detroit Free Press (Jan. 24, 2006) "Icy grip on Russia deadly"
Retrieved 1/28/2006 from:
http://www.freep.com/apps/pbcs.dll/article?AID=/20060124/NEWS07/601240382/1009
CO2 Science (Jan. 4, 2006) "Solar-Powered Millennial-Scale Climatic Change"
Retrieved 1/28/2006 from:
http://www.co2science.org/scripts/CO2ScienceB2C/articles/V9/N1/EDIT.jsp

[4] From FAIR's report: Journalistic Balance as Global Warming Bias
Retrieved 1/27/2006 from:
http://www.fair.org/index.php?page=1978

[5] Washington Post, January 26, 2006, Study: Global Warming May Raise Sea Levels
Retrieved 1/27/2006 from:
http://www.washingtonpost.com/wp-dyn/content/article/2006/01/26/AR2006012601211.html

[6] National Geographic News January 12, 2006 Frog Extinctions Linked to Global Warming
Retrieved 1/27/2006 from:
http://news.nationalgeographic.com/news/2006/01/0112_060112_frog_climate.html

[7] Fox News, January 24, 2006 Study: Global Warming Not Killing Off Arizona Frogs
Retrieved 1/27/2006 from:
http://www.foxnews.com/story/0,2933,182637,00.html

[8] Google News search conducted on 1/28/2006

[9] The book on forecasting is *The WetSand WaveCast Guide to Surf Forecasting: A simple approach to planning the perfect session*

[10] A Short History of Nearly Everything, Bill Bryson, p 255

[11] Callendar, G. S. (1961). *Temperature fluctuations and trends over the Earth*. Quarterly Journal of the Royal Meteorological Society, 87, 112.

[12] Note that the butterfly-effect mentioned here and throughout the book is merely a metaphor to how seemingly small changes can snowball out of control. I realize that the sensitive dependence on initial conditions (e.g. the butterfly effect) actually refers to weather, and not climate. Yet global warming's causality is brought upon by what seems like minor perturbations to our world that later have catastrophic consequences; hence, a butterfly effect *type* of result/condition is created.

[13] Atmospheric Composition, 7a, Dr. Michael Pidwirny, University of British Columbia
Retrieved 2/14/2006 from
http://www.physicalgeography.net/fundamentals/7a.html
Using water as 3%, and CO2 as 0.036% and CH4 as 0.00017%
Note this also includes clouds. If not using clouds, the percentage of water vapor would be only as high as 70% as per:
http://en.wikipedia.org/wiki/Greenhouse_gas
In either case, water vapor is still agreeably the most abundant greenhouse gas around—yet the least damaging and nothing we need to be concerned with.

[14] Global Warming International Center, April 10, 2005 "Researchers highlight the Growing Contribution of Aviation to Global Warming at 16th Global Warming International Conference in New York to coincide with Earthday"
Retrieved 1/30/2006 from:
http://www.globalwarming.net/index.php?option=com_content&task=view&id=117&Itemid=1

[15] Volcanoes only emit around 0.16 billion tons of CO2 annually, according to Environment Canada, Science of Climate Change FAQ
Question B.4
Retrieved 1/28/2006 from:
http://www.msc.ec.gc.ca/education/scienceofclimatechange/understanding/FAQ/sections/2_e.html

[16] National Center for Atmoshperic Research & the UCAR Office of Programs
Retrieved 1/30/2006 from:
http://www.ucar.edu/learn/1_4_2_16t.htm

[17] Environment Canada, Science of Climate Change FAQ
Question B.4
Retrieved 1/28/2006 from:
http://www.msc.ec.gc.ca/education/scienceofclimatechange/understanding/FAQ/sections/2_e.html

[18] Environment Canada, Science of Climate Change FAQ
Excerpt from IPCC, 2001 WGI, Chapter 3
Retrieved 1/28/2006 from:
http://www.msc.ec.gc.ca/education/scienceofclimatechange/understanding/FAQ/sections/2_e.html

[19] U.S. Environmental Protection Agency
Table 1-1 Global Atmospheric Concentrations
Retrieved 1/29/2006 from:
Also see IPCC 2001 The Scientific Basis C.1 Table 1 Observed Changes in Globally Well-Mixed Greenhouse Gas Concentrations and Radiative Forcing
Retrieved 1/29/2006 from:
http://www.grida.no/climate/ipcc_tar/vol4/english/086.htm#co2 and
http://yosemite.epa.gov/oar/globalwarming.nsf/content/ResourceCenterPublicationsGHGEmissionsUSEmissionsInventory2005Tables.html/$File/Table%201-1.csv

[20] U.S. Department of Energy, Energy Information Administration/International Energy Outlook 2005
Retrieved 2/14/2006 from:
http://www.eia.doe.gov/oiaf/ieo/emissions.html

[21] U.S. Environmental Protection Agency
Retrieved 1/30/2006 from:
http://www.epa.gov/methane/

[22] United Exterminating Company of New Jersey
Retrieved 1/30/2006 from:
http://www.unexco.com/Termite.html

[23] U.S. Environmental Protection Agency
Retrieved 1/30/2006 from:
http://www.epa.gov/methane/rlep/faq.html#1

[24] IPCC 2001 The Scientific Basis Chapter 4
Retrieved 1/30/2006 from:
http://www.grida.no/climate/ipcc_tar/wg1/135.htm

[25] U.S. Geological Survey, "Age of the Earth"
Retrieved 1/28/2006 from:
http://pubs.usgs.gov/gip/geotime/age.html

[26] Global Hotmap
Retrieved 1/30/2006 from:
http://www.climatehotmap.org/namerica.html

[27] IPCC Synthesis Report, 2001
Retrieved 1/29/2006 from:
http://www.grida.no/climate/ipcc_tar/vol4/english/008.htm

[28] Wikipedia
Retrieved 3/18/2006 from:
http://en.wikipedia.org/wiki/European_heat_wave_of_2003 and
New Scientist, 10/10/2003, European heat wave causes 35,000 deaths
http://www.newscientist.com/article.ns?id=dn4259

[29] Our existence calculated to less than half an inch compared to one mile, calculated from Earth's age of 4.5 billion years compared to modern humans from 32,000 years ago as per Encarta, where Homo Sapiens first appeared in Europe during the Paleolithic Era.

[30] Health Canada, Drug Bulletin Sep.-2003, item 6
Retrieved 2/11/2006 from:
http://www.hc-sc.gc.ca/fnih-spni/pubs/drug-med/2003-09-bull-lebull/index_e.html

[31] Planet Ark, Arctic Ice and Way of Life Melting for Eskimos
Retrieved 2/11/2006 from:
http://www.climateark.org/articles/reader.asp?linkid=9488

[32] Anchorage Daily News, Alaska villagers say climate change is forcing community changes
Retrieved 2/11/2006 from:
http://www.adn.com/news/alaska/ap_alaska/story/7427126p-7337979c.html

[33] Reuters 12/2/2005, Arctic feels the heat from climate change
Retrieved 12/6/2005 from:
http://www.alertnet.org/thenews/newsdesk/SP137755.htm

[34] CBS News, Northern hunters worry about U.S. polar bear decision
Retrieved 2/11/2006 from:
http://www.cbc.ca/story/canada/national/2006/02/09/polarbear-us060209.html

[35] SFGate, Polar bears to be considered for threatened species list
Retrieved 2/11/2006 from:
http://www.sfgate.com/cgi-bin/article.cgi?file=/c/a/2006/02/09/MNGH6H58GK1.DTL

[36] Big Sur Chamber of Commerce, originally reported in the LA Times June 24, 2002, A Whale of a Food Shortage
Retrieved 2/15/2006 from:
http://www.bigsurcalifornia.org/whalesgray-LATimes6.24.02.html

[37] National Geographic News, 1/18/2006 Whale Birth Decline Tied to Global Warming, Study Says
Retrieved 2/15/2006 from:
http://news.nationalgeographic.com/news/2006/01/0118_060118_right_whales.html
And also from San Francisco Chronicle, SFGate, 2/19/2005, New global warming evidence presented: Scientists say their observations prove industry is to blame
Retrieved 2/15/2006 from:
http://sfgate.com/cgi-bin/article.cgi?file=/c/a/2005/02/19/MNGE1BECPI1.DTL

[38] BBC News, Global warming helps Arctic animals 11, May 2001
Retrieved 2/15/2006 from:
http://news.bbc.co.uk/1/hi/world/americas/1324416.stm
Also note that the reference in this article to increased polar bear populations has no link to climate change, but instead mentions it is due to a hunting ban.

[39] Anchorage Daily News, Alaska villagers say climate change is forcing community changes
Retrieved 2/11/2006 from:
http://www.adn.com/news/alaska/ap_alaska/story/7427126p-7337979c.html

[40] The Tyee, 1/30/2006, The Need to Defend Our New Northwest Passage
Retrieved 2/15/2006 from:
http://thetyee.ca/Views/2006/01/30/DefendNorthwestPassage/

[41] Genetic shift in photoperiodic response correlated with global warming William E. Bradshaw
and Christina M. Holzapfel, Nov. 6, 2001
Retrieved 2/15/2006 from:
http://www.pnas.org/cgi/content/full/241391498v1

[42] Inuit Tapiriit Kanatami Environment Bulletin No. 1, 2005, and No. 2, 2005

[43] Grist magazine, 7/26/2005, The Snow Must Go On
Retrieved 215/2006 from:
http://www.grist.org/news/maindish/2005/07/26/gertz-inuit/

[44] Inuit Circumpolar Conference, Dec. 7, 2005, Inuit Petition Inter-American Commission on
Human Rights to Oppose Climate Change Caused by the United States of America
Retrieved 2/15/2006 from:
http://www.inuitcircumpolar.com/index.php?ID=316&Lang=En

[45] BBC News, 6/14/1999, Islands disappear under rising seas
Retrieved 1/24/2006 from:
http://news.bbc.co.uk/1/hi/sci/tech/368892.stm

[46] UNEP, 12/12/2005, Pacific Island Villagers First Climate Change "Refugees"
Retrieved 2/16/2006 from:
http://www.grida.no/newsroom.cfm?pressReleaseItemID-978

[47] ABC News 12/5/2005, Pacific islanders move to escape global warming
Retrieved 2/18/2006 from:
http://abcnews.go.com/US/wireStory?id=1374770

[48] ABC News in Science, 5/28/2003, Sinking atolls trigger Papuan evacuation plans
Retrieved 2/18/2006 from:
http://www.abc.net.au/science/news/stories/s866600.htm

[49] Global Warming: The Complete Briefing, John Houghton (Cambridge Press) p.146

[50] Global Warming: The Complete Briefing, John Houghton (Cambridge Press) p.145

[51] CNN, 2/16/2006, Greenland glaciers dumping ice into Atlantic at faster pace
Retrieved 2/16/2006 from:
http://www.cnn.com/2006/TECH/science/02/16/greenland.glaciers.ap/index.html
Also from NASA, 12/01/04 Fastest Glacier in Greenland Doubles Speed
Retrieved 2/18/2006 from:
http://www.nasa.gov/vision/Earth/lookingatEarth/jakobshavn.html

[52] MSNBC, 2/9/2006, Wrapping glaciers…are they serious?
Retrieved 2/18/2006 from:
http://www.msnbc.msn.com/id/11256540/

[53] GeoTimes, July 2005, Swiss wrap glacier for summer
Retrieved 2/18/2006 from:
http://www.geotimes.org/july05/NN_glacierwrap.html

[54] Global Warming: The Complete Briefing, John Houghton (Cambridge Press) p.147

[55] CNews, 2/14/2006, Arctic climate study produces startling findings
Retrieved 2/15/2006 from:
http://cnews.canoe.ca/CNEWS.Sceince/2006/02/14/pf-1442809.html

[56] State of Florida Quick Facts
Retrieved 2/19/2006 from:
http://www.stateofflorida.com/Portal/DesktopDefault.aspx?tabid=95

[57] U.S. Census Bureau Data set R2510, 2004 American Community Survey
Retrieved 2/10/2006 from:
http://factfinder.census.gov/servlet/GRTTable?_bm=y&-geo_id=01000US&-_box_head_nbr=R2510&-ds_name=ACS_2004_EST_G00_&-format=US-30

[58] U.S. Census Bureau
Retrieved 2/19/2006 from:
http://quickfacts.census.gov/qfd/states/12000.html

[59] Natural Resources Defense Council, 10/23/2001, Florida Scientists Warn Global Warming Spells Trouble for State Economy

Retrieved 2/19/2006 from:
http://www.nrdc.org/media/pressReleases/011023.asp

[60] U.S. EPA
Retrieved 2/19/2006 from:
http://yosemite.epa.gov/OAR/globalwarming.nsf/content/ImpactsCoastalZonesSouthFlorida2.ht
ml

[61] U.S. EPA
Retrieved 2/19/2006 from:
http://yosemite.epa.gov/OAR/globalwarming.nsf/content/ImpactsCoastalZonesSouthFlorida2.ht
ml
and Global Warming: The Complete Briefing, John Houghton (Cambridge Press) p.150

[62] National Hurricane Center
Retrieved 2/19/2006 from:
http://www.nhc.noaa.gov/Deadliest_Costliest.shtml

[63] NOAA Fisheries
Retrieved 2/20/2006 from:
http://www.nmfs.noaa.gov/pr/species/invertebrates/elkhorn.htm

[64] State of Florida.com
Retrieved 2/20/2006 from:
http://www.stateofflorida.com/Portal/DesktopDefault.aspx?tabid=95

[65] NOAA's Coral Health and Monitoring Center
Retrieved 2/20/2006 from:
http://www.coral.noaa.gov/cleo/coral_bleaching.shtml

[66] U.S. EPA Global Warming Impacts
Retrieved 2/20/2006 from:
http://yosemite.epa.gov/oar/globalwarming.nsf/content/ImpactsCoastalZonesChesapeakeBay.html

[67] NOAA, National Geodetic Survey, Variations in Sea Level
Retrieved 2/20/2006 from:
http://www.ngs.noaa.gov/GRD/GPS/Projects/CB/SEALEVEL/sealevel.html

[68] Chesapeake Bay Program, Habitat Preservation and Restoration
Retrieved 2/21/206 from:
http://www.chesapeakebay.net/restrtn.htm

[69] Chesapeake Bay Program, Habitat Preservation and Restoration
Retrieved 2/21/206 from:
http://www.chesapeakebay.net/newsoystersstress090905.htm

[70] Chesapeake Bay Program, Habitat Preservation and Restoration
Retrieved 2/21/206 from:
http://www.chesapeakebay.net/land.htm

[71] Washington Post, 2/6/2005, Arid Arizona Points to Global Warming as Culprit
Retrieved 2/21/2006 from:
http://www.washingtonpost.com/wp-dyn/articles/A1493-2005Feb5.html

[72] Southwest Farm Press, 2/10/2006, Rain puddles add little to Southwest drought relief
Retrieved 2/21/2006 from:
http://southwestfarmpress.com/news/06-02-10-rain-drought-relief/

[73] U.S. Geological Survey, Major Floods and Droughts of Arizona
Retrieved 2/21/2006 from:
http://geochange.er.usgs.gov/sw/impacts/hydrology/state_fd/azwater1.html

[74] USDA, U.S. Drought Monitor 2/14/2006
Retrieved 2/21/2006 from:
http://drought.unl.edu/dm/monitor.html

[75] NOAA, NCDC Climate of 2006–January Arizona Moisture Status 2/3/2006
Retrieved 2/21/2006 from:
http://www.ncdc.noaa.gov/oa/climate/research/prelim/drought/st002dv00pcp.html

[76] NOAA, NCDC Climate of 2006–January in Historical Perspective 2/14/2006
Retrieved 2/21/2006 from:
http://lwf.ncdc.noaa.gov/oa/climate/research/2006/jan/jan06.html

[77] NASA 1/24/2006, 2005 Was the Hottest Year in a Century
Retrieved 1/25/2006 from:
http://www.nasa.gov/vision/Earth/environment/2005_warmest.html
Also Time magazine, February 6, 2006 p18
And CNN.com 1/24/2006 2005 was the hottest year in a century
Retrieved 2/21/2006 from:
http://www.cnn.com/2006/WEATHER/01/24/hot.year.ap/

[78] National Geographic News, 2/9/2006 Current Warming Period Is Longest in 1,200 Years, Study Says
Retrieved 2/10/2006 from:
http://news.nationalgeographic.com/news/2006/02/0209_060209_warming.html

[79] NOAA, NCDC Global Hazards and Significant Events, January 2006
Retrieved 2/21/2006 from:
http://lwf.ncdc.noaa.gov/oa/climate/research/2006/jan/hazards.html#Drought

[80] USGS, From the Dust Bin of History to a Future of Integrated Natural Science By Dale Griffin and Gene Shinn
Retrieved 3/9/2006 from:
http://www.usgs.gov/125/articles/dust.html
New Scientist.com 8/20/2004 Dust storms on the rise globally
Retrieved 2/10/2006 from:
http://www.newscientist.com/article.ns?id=dn6306&print=true

[81] Geo Times, June 2004
Retrieved 2/10/2006 from:
http://www.geotimes.org/june04/geophen.html

[82] Jerusalem Post 2/9/2006 Dust storms now hit Israel year round

[83] USGS Coral Mortality and African Dust
Retrieved 2/10/2006 from:
http://coastal.er.usgs.gov/african_dust

[84] Berkley Lab Research News 10/24/2002 Asian dust storm causes plankton to bloom in the North Pacific
Retrieved 2/10/2006 from:
http://www.lbl.gov/Science-Articles/Archives/ESD-Gobi-plankton-Bishop.html

[85] BBC News 6/30/2000 Chinese gourmets destroy desert
Retrieved 2/15/2006 from:
http://news.bbc.co.uk/1/hi/word/asia-pacific/813254.stm

[86] National Geographic News, 6/1/2001, China's Dust Storms Raise Fears of Impending Catastrophe
Retrieved 2/14/2006 from:
http://news.nationalgeographic.com/news/2001/06/0601_chinadust.html

[87] Scientific American, 6/9/2003, Disease Dustup
Retrieved 2/23/2006 from:
http://www.sciam.com/article.cfm?articleID=000ECFD9-15AB-1EE1-A2D1809EC5880000

[88] Climate Hotmap
Retrieved 2/22/2006 from:
http://www.climatehotmap.org/camerica.html

[89] University of Wisconsin-Madison, 11/16/2005, Third World bears brunt of global warming impacts
Retrieved 2/22/2006 from:
http://www.news.wisc.edu/11878.html

[90] Reuters, 12/5/2005, Europe's health woes may worsen with global warming
Retrieved 12/6/2005 from:
http://www.alertnet.org/thenews/newsdesk/N05254496.htm

[91] CDC 2000 and 2005 West Nile Virus Activity in the United States
Retrieved 2/22/2006 from:
http://www.cdc.gov/ncidod/dvbid/westnile/surv&controlCaseCount05_detailed.htm
and http://www.cdc.gov/ncidod/dvbid/westnile/surv&controlCaseCount00_detailed.htm

[92] Japan Agrinfo Newsletter, the Japan Association for International Collaboration of Agriculture and Forestry
Retrieved 2/22/2006 from:
http://www.jaicaf.or.jp/agrinfo/0511/Report_230306.htm

[93] CNet News, "Global warming to help you breath?" 2/10/2006 by Michael Kanellos, refers to a study by Gavin Donaldson from University College London
Retrieved 4/22/2006 from:
http://news.com.com/Global+warming+to+help+you+breathe/2100-11393_3-6037939.html

[94] Hurricane information data and states retrieved from NOAA NCDC:
Retrieved 2/24/2006 from:
http://www.ncdc.noaa.gov/oa/climate/research/2005/hurricanes05.html

[95] NOAA
Retrieved 2/25/2006 from:
http://www.srh.weather.gov/jetstream/tropics/tip.htm

[96] NOAA News Online Story 2540b, Nov. 29, 2005
Retrieved 2/25/2006 from:
http://www.noaanews.noaa.gov/stories2005/s2540b.htm

[97] Sea surface temperatures for the North Atlantic from NOAA NCEP
Retrieved 2/24/2006 from:
http://www.cpc.ncep.noaa.gov/data/indices/sstoi.atl.indices

[98] National Center for Policy Analysis (NCPA), Hot Air vs. the Cold Hard Truth about Hurricanes and Global Warming
Retrieved 2/25/2006 from:
http://www.ncpa.org/pub/ba/ba530/

[99] San Francisco State University, Dept. of Geosciences
Retrieved 2/25/2006 from:
http://tornado.sfsu.edu/geosciences/classes/m302/stability.html
and http://tornado.sfsu.edu/geosciences/classes/m356/Instability.html

[100] George Mason University, Menas Kafatos, et al., Anomalous Gulf Heating and Hurricane Katrina's Rapid Intensification
Retrieved 2/25/2006 from:
http://arxiv.org/ftp/physics/papers/0509/0509177.pdf

[101] NASA Earth Observatory, Hurricane Alley Heats Up
Retrieved 2/25/2006 from:
http://Earthobservatory.nasa.gov/Newsroom/NewImages/images.php3?img_id=16996

[102] NOAA, Billy Kessler, Pacific Marine Environment Laboratory
Retrieved 2/25/2006 from:
http://www.pmel.noaa.gov/%7Ekessler/occasionally-asked-questions.html

[103] NOAA NWS, CPC, Cold & Warm Episodes by Season
Retrieved 2/26/2006 from:
http://www.cpc.ncep.noaa.gov/products/analysis_monitoring/ensostuff/ensoyears.shtml

[104] NOAA, Billy Kessler, Pacific Marine Environment Laboratory, FAQ 3
Retrieved 2/25/2006 from:
http://www.pmel.noaa.gov/%7Ekessler/occasionally-asked-questions.html

[105] National Hurricane Center, NCEP archive
Retrieved 2/26/2006 from:
http://www.nhc.noaa.gov/1997epac.html

[106] National Geographic Magazine, El Nino/La Nina Nature's Vicious Cycle
Retrieved 2/26/2006 from:
http://www.nationalgeographic.com/elnino/mainpage.html
Also Golden Gate Weather Services from:
http://ggweather.com/nino/calif_flood.html

[107] CISRO Climate Change Research Program 1995-1999 Summary Report November 1999
Retrieved 2/26/2006 from:
http://www.cmar.csiro.au/ar/csiro_reserved/CCRP99/CCRP%201995_99_outcomes_6.pdf
Also the Australian Government Department of the Environment and Heritage, Australian
Greenhouse Office, Fact Sheet March 1999, Retrieved 2/26/2006 from:
http://www.greenhouse.gov.au/science/factsheet/enso.html

[108] BBC News 10/25/2000, Coral record reveals secrets of El Nino
Retrieved 2/26/2006 from:
http://news.bbc.co.uk/1/hi/sci/tech/990670.stm
Article refers to studies by Scientists at the University of Colorado, Boulder, studied a core of 155-
year-old coral drilled at Maiana Atoll in the central Pacific.
Additionally, there were studies conducted by Dr. Jerry Wellington of the University of Houston,
and Dr. Rob Dunbar of Stanford University in the Galapagos in 1997, see report by NPR
retrieved 2/26/2006 from:
http://www.npr.org/programs/re/archivesdate/1998/aug/19980810.galapagos.html

[109] IPCC TAR 2001 The Scientific Basis, Summary for Policymakers, p. 5

[110] PDO time cycles were derived from averages taken from the PDO index data, 2/27/2006
from:
http://jisao.washington.edu/pdo/PDO.latest
This was coincided with charts at:
http://tao.atmos.washington.edu/pdo/

[111] Based on U.S. Hurricane Strikes per Decade from the National Hurricane Center
Retrieved 2/26/2006 from:
http://www.nhc.noaa.gov/pastdec.shtml

[112] NOAA, AOML, Frequently Asked Questions about the AMO
Retrieved 2/27/2006 from:
http://www.aoml.noaa.gov/phod/amo_faq.php#faq_6

[113] Census stats for Florida derived from U.S. Census Bureau and Florida Legislature's EDR Program
Retrieved 2/28/2006 from:
http://quickfacts.census.gov/qfd/states/12000.html
and http://edr.state.fl.us/population/table1-4.xls

[114] National Geographic, August 2005, In Hot Water
Retrieved 2/28/2006 from:
http://www7.nationalgeographic.com/ngm/0508/feature4/

[115] Planktonic Forminifera of the California Current Reflect 20th Century Warming, David B. Field, et al.
Also referenced from news article from MBARI news, retrieved 2/28/2006 from:
http://www.mbari.org/news/news_releases/2006/forams.html

[116] Key History, History of the Gulf Stream, By Jerry Wilkinson
Retrieved 3/1/2006 from:
http://www.keyshistory.org/gulfstream.html

[117] National Geographic, August 2005, In Hot Water
Retrieved 2/28/2006 from:
http://www7.nationalgeographic.com/ngm/0508/feature4/
Also from Gray's hurricane forecast from:
http://hurricane.atmos.colostate.edu/forecasts/2005/april2005/
Predictor 4, re: above normal SSTs off the Northwestern European Coast reflecting a likely stronger than normal

[118] Wikipedia, Younger Dryas
Retrieved 2/28/2006 from:
http://en.wikipedia.org/wiki/Younger_Dryas

[119] BBC h2g2 Younger Dryas, the Ice Age's Last Big Blast, A760240
Retrieved 3/1/2006 from:
http://www.bbc.co.uk/dna/h2g2/plain/A760240

[120] Nature 438, 655-657 (1 December 2005) | doi:10.1038/nature04385 "Slowing of the Atlantic meridional overturning circulation at 25° N" Harry L. Bryden1 et al
Retrieved 3/20/2006 from:
http://www.nature.com/nature/journal/v438/n7068/abs/nature04385.html
Also see story from Independent Online "Gulf Stream 'engine' weakening, say scientists" By John von Radowitz, PA Published: 30 November 2005 from:
http://news.independent.co.uk/environment/article330361.ece

[121] Crichton's Senate Speech was retrieved 3/4/2006 from:
http://www.crichton-official.com/speeches/senate.html

[122] The Guardian, 9/29/2005, Novel take on global warming By Jamie Wilson
Retrieved 12/16/2005 from:
http://www.guardian.co.uk/print/0,3858,5296880-103677,00.html

[123] Inhofe speech retrieved 3/4/2006 from:
http://inhofe.senate.gov/pressreleases/climateupdate.htm

[124] Inhofe speech retrieved 7/28/2003 from:
http://inhofe.senate.gov/pressreleases/climate.htm

[125] Taken from a transcript from the Hearing Before the Committee on Environment and Public Works, United States Senate, 108th Congress, First Session, July 29, 2003
Retrieved 3/3/2006 from:
http://frwebgate.access.gpo.gov/cgi-bin/getdoc.cgi?dbname=108_senate_hearings&docid=f:92381.wais

[126] Study refers to Soon and Baliunas, "Proxy climatic and environmental changes of the past 1000 years" published in Climate Research, CR 23:89-110(2003), ISSN: 0936-577X

[127] Michael Mann's bio excerpt from Wikipedia
Retrieved 3/4/2006 from:
http://en.wikipedia.org/wiki/Michael_Mann_(scientist)

[128] Science, Vol 307 2/11/2005, p828

[129] Australian Government Dept. of the Environment and Heritage, Australian Greenhouse Office, "hot topics in climate change science", "How Unusual is the Late 20th Century Warming", topic 3, April 2005, page 2, referring to Mann et al 2003b

[130] The M&M Critique of the MBH98 Northern Hemisphere Climate Index: Update and Implications, Stephen McIntyre, Ross McKitrick, the journal Energy and Environment Vol. 16 No1. 2005

[131] The M&M Critique of the MBH98 Northern Hemisphere Climate Index: Update and Implications, Stephen McIntyre, Ross McKitrick, the journal Energy and Environment Vol. 16 No1. 2005 Figure 1
Also from World Climate Report, referring to Esper study, retrieved 3/4/2006 from:
http://www.worldclimatereport.com/index.php/2005/03/03/hockey-stick-1998-2005-rip/
See Figure 2.
Third reference from Science, Vol 307 2/11/2005, p828, see graph representing Overpeak97, Jones98, Mann99, Briffa00, Briff01, Esper02,Obs, and Moberg05

[132] Hypertextbook.com
Retrieved 3/4/2006 from:
http://hypertextbook.com/facts/2000/IlanaPrice.shtml

[133] Oak Ridge National Laboratories, Environmental Sciences Division, "A quick background to the last ice age"
Retrieved 3/2/2006 from:
http://www.esd.ornl.gov/projects/qen/nerc130k.html

[134] Wikipedia, Eemian
Retrieved 3/8/2006 from:
http://en.wikipedia.org/wiki/Eemian

[135] IPCC 2001 WG1, The Scientific Basis 2.4.2
Retrieved 3/4/2006 from:
http://www.grida.no/climate/ipcc_tar/wg1/073.htm and also from Wikipedia, Holocene climatic optimum
Retrieved 3/4/2006 from: http://en.wikipedia.org/wiki/Holocene_Climatic_Optimum

[136] Wikipedia
Retrieved 3/5/2006 from:
http://en.wikipedia.org/wiki/Sun and also from Hpertextbook from
http://hypertextbook.com/facts/2000/CCoraThomas.shtml

[137] Solar Variability and Global Climatic Change Sallie Baliunas and Willie Soon
Retrieved 3/5/2006 from:
http://oldfraser.lexi.net/publications/books/g_warming/solar.html

[138] Raimund Muscheler, commenting on RealClimate, 8/3/2005, "Did the Sun hit record highs over the last few decades?"
Retrieved 3/5/2006 from:
http://www.realclimate.org/index.php?p=180#more-180

[139] Milankovitch Cycles and Glaciation, Montana State University, Geology 445, W.W. Locke
Retrieved 3/5/2006 from:
http://www.homepage.montana.edu/~geol445/hyperglac/time1/milankov.htm

[140] Messier Objects, Chris Dolan, Dept. of Astronomy University of Wisconsin-Madison
Retrieved 3/5/2006 from:
http://www.astro.wisc.edu/~dolan/constellations/constellations/Lyra.html

[141] Global Warming, The Complete Briefing, John Houghton p. 70

[142] San Diego State University Dept. of Geological Sciences, Climate Effects of Volcanic Eruptions
Retrieved 3/2./006 from:
http://www.geology.sdsu.edu/how_volcanoes_work/climate_effects.html also from USGS site at:
http://vulcan.wr.usgs.gov/Volcanoes/Indonesia/description_krakatau_1883_eruption.html

[143] University of North Dakota's Volcano World, "What happened in the Krakatoa eruption in the 1800's?" by Nicholas Elliott, retrieved 6/26,2006 from:
http://volcano.und.edu/vwdocs/frequent_questions/grp7/asia/question879.html

[144] Environment Canada, Science of Climate Change FAQ Question B.4
Retrieved 1/28/2006 from:
http://www.msc.ec.gc.ca/education/scienceofclimatechange/understanding/FAQ/sections/2_e.html

[145] San Diego State University Dept. of Geological Sciences, Climate Effects of Volcanic Eruptions
Retrieved 3/2./006 from:
http://www.geology.sdsu.edu/how_volcanoes_work/climate_effects.html also from Global Warming, The Complete Briefing, John Houghton p. 8

[146] University of North Dakota, Volcano World, Laki, Iceland–1783
Retrieved 3/5/2006 from:
http://volcano.und.edu/vwdocs/Gases/laki.html

[147] University of North Dakota, Steve Mattox, "Was the Tambora eruption of 1815 responsible for 'the year without a summer' in 1816?"

Retrieved 3/30/2006 from:

http://volcano.und.nodak.edu/vwdocs/frequent_questions/grp4/question1195.html

And BBC h2g2 "Tambora—the Greatest Volcanic Eruption in Recorded History" 29th July 2002 from:

http://www.bbc.co.uk/dna/h2g2/A781715

[148] University of California San Diego, "The Tambora Eruption: Historical Accounts of the Event"

Retrieved 3/30/2006 from:

http://earthguide.ucsd.edu/sun/tambora.html

[149] U.S. EPA, "Greenhouse Gases and Global Warming Potentials" April 2002, p. 9, Table 2

Retrieved 3/5/2006 from:

http://yosemite.epa.gov/oar/globalwarming.nsf/UniqueKeyLookup/SHSU5BUM9T/$File/ghg_gwp.pdf

[150] Lawrence Livermore National Laboratories, News Release, "Volcanoes helped slow ocean warming trend, researchers find"

Retrieved 3/2/2006 from:

http://www.llnl.gov/pao/news/news_releases/2006/NR-06-02-02p.html

[151] Wall Street Journal, OpinionJournal, 6/11/2001, "The Press Gets it Wrong", Richard S. Lindzen

Retrieved 3/6/2006 from:

http://opinionjournal.com/editorial/feature.html?id=95000606

[152] IPCC SAR Working Group 1

Retrieved 3/6/2006 from:

http://www.ipcc.ch/pub/sarsum1.htm

[153] IPCC TAR Working Group 1

Retrieved 3/6/2006 from:

http://www.grida.no/climate/ipcc_tar/wg1/007.htm

[154] Science, Vol. 306 no. 5702, p.1686 DOI: 10.1126/science.1103618, "Beyond the Ivory Tower, The Scientific Consensus on Climate Change" Naomi Oreskes

[155] National Academy of Sciences Committee on the Science of Climate Change, Climate Change Science: An Analysis of Some Key Questions (National Academy Press, Washington, DC, 2001)

[156] American Meteorological Society, Climate Change Research: Issues for the Atmospheric and Related Sciences (Adopted by AMS Council on 9 February 2003) Bull. Amer. Met. Soc., 84, 508—515
Retrieved 3/6/2006 from:
http://www.ametsoc.org/policy/climatechangeresearch_2003.html

[157] Gulf of Maine Research Institute, Antarctica
Retrieved 3/6/2006 from:
http://octopus.gma.org/surfing/antarctica/antarctica.html

[158] Drilling sites refers to EPICA cores taken in 2005.

[159] Carbon Dioxide Information Analysis Center (U.S. DOE), "Historical carbon dioxide record from the Vostok ice core", J.-M. Barnola, D. Raynaud, C. Lorius
Retrieved 3/23/2006 from:
http://cdiac.ornl.gov/trends/co2/vostok.htm

[160] Science, VOL 299 3/14/2003, Timing of Atmospheric CO_2 and Antarctic Temperature Changes Across Termination III, Caillon, et. al.

[161] IPCC 2001 Synthesis Report, question 2.4
Retrieved 1/30/2006 from:
http://www.grida.no/climate/ipcc_tar/vol4/english/019.htm

[162] Science, Sciencexpress, 3/2/2006, Page 1, 10.1126/science.1123785, Isabella Velicogna and John Wahr, "Measurements of Time-Vriable Gravity Show Mass Loss in Antarctica"

[163] IPCC, Principles Governing IPCC Work Nov. 2003
Retrieved 3/10/2006 from:
http://www.ipcc.ch/about/princ.pdf

[164] TSAugust, Climate Scientist Chris Landsea Quits IPCC
Retrieved 3/10/2006 from:
http://www.tsaugust.org/Scientists%20are%20Saying.htm and from University of Colorado, Prometheus, 1/17/2005, Chris Landsea Leaves IPCC
Retrieved 3/10/2006 from:
http://sciencepolicy.colorado.edu/prometheus/archives/science_policy_general/000318chris_land-sea_leaves.html

[165] American Scientist Online, May-June 2005, A tempest erupts over the political neutrality of the best-known climate-change panel by David Schneider
Retrieved 3/10/2006 from:
http://www.americanscientist.org/template/AssetDetail/assetid/42399

[166] Email from Chris Landsea to various members of the IPCC, 11/4/2005, final paragraph
Retrieved 3/10/2006 from:
http://sciencepolicy.colorado.edu/prometheus/archives/ipcc-correspondence.pdf

[167] University of Colorado, Prometheus, 1/17/2005, Chris Landsea Leaves IPCC
Retrieved 3/10/2006 from:
http://sciencepolicy.colorado.edu/prometheus/archives/science_policy_general/000318chris_landsea_leaves.html

[168] Nature, v.436 4aug2005, Increasing destructiveness of tropical cyclones over the past 30 years, Kerry Emanuel

[169] Bulletin of the American Meteorological Society, Nov. 2005, p.1571, "Hurricanes and Global Warming" by R.A. Pielke Jr., C. Landsea, et al.

[170] Nature Vol 438 22/29 Dec. 2005 pE11, "Are there trends in hurricane destruction?" by Roger Al. Pielke Jr. and "Hurricanes and global warming" by Christopher W. Landsea

[171] Science Vol. 309 no. 5742 pp. 1844-1846 9/16/2005, "Changes in Tropical Cyclone Number, Duration, and intensity in a Warming Environment" by P.J. Webster, et al.

[172] National Geographic News, 9/15/2005, "Hurricanes are Getting Stronger, Study Says" by John Roach
Retrieved 3/11/2006 from:
http://news.nationalgeographic.com/news/2005/09/0915_050915_hurricane_strength.html

[173] American Institute of Biological Sciences, "AIBS Public Policy Report for 11 October 2005"
Retrieved 3/11/2006 from:
http://www.aibs.org/public-policy-reports/public-policy-reports-2005_10_11.html see section titled "Hollywood comes to DC: Michael Crichton testifies about climate change in the Senate"

[174] San Diego Union Tribune, "Global warming's link to storm power debated" 9/14/2006 by Bruce Lieberman
Retrieved 2/25/2006 from:
http://www.signonsandiego.com/uniontrib/20050914/news_lz1c14link.html

[175] The Tropical Meteorology Project, Colorado State University "EXTENDED RANGE FORE-CAST OF ATLANTIC SEASONAL HURRICANE ACTIVITY AND U.S. LANDFALL STRIKE PROBABILITY FOR 2006"12/6/2005 By Philip J. Klotzbach and William M. Gray

[176] The National Academies, Joint science academies' statement: Global response to climate change, 6/7/2005
Retrieved 3/11/2006 from:
http://nationalacademies.org/onpi/06072005.pdf

[177] "Climate Change Science, An Analysis of Some Key Questions" National Academy Press, Washington, D.C. 2001
Retrieved 3/11/2006 from:

[178] American Meteorological Society, Executive Summary, Climate Change Research: Issues for the Atmospheric and Related Sciences (Adopted by AMS Council on 9 February 2003) Bull. Amer. Met. Soc., 84, 508—515
Retrieved 3/11/2006 from:
http://www.ametsoc.org/policy/climatechangeresearch_2003.html

[179] Science 12/3/2004 Vol 306 no. 5702 p 1686, "BEYOND THE IVORY TOWER: The Scientific Consensus on Climate Change", Naomi Oreskes
Retrieved 3/11/2006 from:
http://www.sciencemag.org/cgi/content/full/306/5702/1686

[180] Wikipedia "Decline in Frog Populations"
Retrieved 3/12/2006 from:
http://en.wikipedia.org/wiki/Decline_in_frog_populations

[181] Nature, Nature 439, 161-167 (12 January 2006) | doi:10.1038/nature04246 "Widespread amphibian extinctions from epidemic disease driven by global warming"
J. Alan Pounds et. al
Also extrapolated from National Geographic News, "Frog Extinctions Linked to Global Warming" by
Brian Handwerk January 12, 2006
Retrieved 1/19/2006 from:
http://news.nationalgeographic.com/news/2006/01/0112_060112_frog_climate.html

[182] IPCC TAR 2001 Summary for Policymakers, p 42

[183] IPCC TAR 2001 Summary for Policymakers, p 38 table 1

[184] IPCC TAR 2001 Working Group II Impacts, Adaptation, and Vulnerability 16.2.5.2

[185] NewScientist.com "Climate warning as Siberia melts" 11 August 2005 Fred Pearce
Retrieved 3/29/2006 from:
http://www.newscientist.com/channel/earth/mg18725124.500

[186] IPCC TAR 2001 Summary for Policymakers, p 67 F.3

[187] Environment Canada, Science of Climate Change, North American Projections, Temperature
Retrieved 3/23/2006 from:
http://www.msc-
smc.ec.gc.ca/education/scienceofclimatechange/understanding/climate_models/index_e.html

[188] Global Warming: The Complete Briefing, John Houghton (Cambridge Press) p.146

[189] Global Warming: The Complete Briefing, John Houghton (Cambridge Press) p.145

[190] Global Warming: The Complete Briefing, John Houghton (Cambridge Press) p.150

[191] IPCC TAR 2001 Summary for Policymakers, p 77

[192] Science, Sciencexpress, 3/2/2006, Page 1, 10.1126/science.1123785, Isabella Velicogna and
John Wahr, "Measurements of Time-Vriable Gravity Show Mass Loss in Antarctica"

[193] Global Warming the Complete Briefing, John Houghton, p147

[194] IPCC TAR 2001 Summary for Policymakers, p 77

[195] City Data, retrieved 3/14/2006 from:
http://www.city-data.com/city/Miami-Beach-Florida.html

[196] U.S. EPA, Global Warming Impacts, Coastal Zones
Retrieved 3/154/2006 from:
http://yosemite.epa.gov/oar/globalwarming.nsf/content/ImpactsCoastalZones.html

[197] FEMA, 6/27/2000 Significant Losses From Coastal Erosion Anticipated Along U.S. Coastlines
Retrieved 3/14/2006 from:
http://www.fema.gov/nwz00/erosion.shtm

[198] FEMA report published at EPA website, Projected Impact of Relative Sea Level Rise on the National Flood Insurance Program
Retrieved 3/14/2006 from:
http://yosemite.epa.gov/oar/globalwarming.nsf/content/ResourceCenterPublicationsSLRFlood_Insurance.html

[199] EPA Brochure, Climate Change and Cold Water Fish, May 1999
Retrieved 3/14/2006 from:
http://yosemite.epa.gov/OAR/globalwarming.nsf/UniqueKeyLookup/SHSU5BNNWD/$File/ccandcoldwaterfish.pdf

[200] EPA, Ecological Impacts From Climate Change: An Economic Analysis of Freshwater Recreational Fishing, U.S. Environmental Protection Agency Report April 1995, EPA #220-R-95-004
Retrieved 3/14/2006 from:
http://yosemite.epa.gov/OAR/globalwarming.nsf/content/ResourceCenterPublicationsEco_fishing.html

[201] EPA, Global Warming Impacts, Fisheries, Inland
Retrieved 3/14/2006 from:
http://yosemite.epa.gov/OAR/globalwarming.nsf/content/ImpactsFisheriesInland.html

[202] LaCoast.gov, Watermarks 2006, ""

[203] EPA, Global Warming Impacts, Fisheries, Coastal
Retrieved 3/14/2006 from:
http://yosemite.epa.gov/OAR/globalwarming.nsf/content/ImpactsFisheriesCoastal.html

[204] RealClimate, CO2 Fertilization, 11/28/2004
Retrieved 3/16/2006 from:
http://www.realclimate.org/index.php?p=93 also from FACE Program, Brookhaven National Laboratory from: http://www.face.bnl.gov/face1.htm

[205] Norby et al., 1999, per IPCC TAR, chapter 5, page 294
Retrieved 3/16/2006 from:
http://www.grida.no/climate/ipcc_tar/wg2/pdf/wg2TARchap5.pdf

[206] IPCC TAR chapter 5, page 290
Retrieved 3/16/2006 from:
http://www.grida.no/climate/ipcc_tar/wg2/pdf/wg2TARchap5.pdf

[207] EPA Global Warming Impacts, Forests
Retrieved 3/16/2006 from:
http://yosemite.epa.gov/oar/globalwarming.nsf/content/ImpactsForests.html

[208] IPCC TAR chapter 5, page 294
Retrieved 3/16/2006 from:
http://www.grida.no/climate/ipcc_tar/wg2/pdf/wg2TARchap5.pdf

[209] IPCC TAR chapter 5, page 290
Retrieved 3/16/2006 from:
http://www.grida.no/climate/ipcc_tar/wg2/pdf/wg2TARchap5.pdf

[210] Rangelands West, What are Rangelands
Retrieved 3/16/2006 from:
http://rangelandswest.org/whatarerangelands.html

[211] IPCC TAR 2001, 5.5.1
Retrieved 3/16/2006 from:
http://www.grida.no/climate/ipcc_tar/wg2/232.htm

[212] Oregon State University, Dr. Patricia Muir, Western Public Lands as an Example
Retrieved 3/17/2006 from:
http://oregonstate.edu/~muirp/wpubland.htm

[213] Oregon State University, Dr. Patricia Muir, Western Public Lands as an Example
Retrieved 3/17/2006 from:
http://oregonstate.edu/~muirp/wpubland.htm

[214] Society for Range Management, Rangelands and Global Change
Retrieved 3/17/2006 from:
http://www.rangelands.org/pdf/Global_Issue_Paper.pdf

[215] U.S. Fish and Wildlife Service, Mountain–Prairie Region, Endangered Species Program,
Greater Sage-Grouse
Retrieved 3/17/2006 from:
http://mountain-prairie.fws.gov/species/birds/sagegrouse/

[216] Pew Center on Global Climate Change, A review of impacts to U.S. agricultural resources, by
Richard M. Adams Oregon State University, Brian H. Hurd Stratus Consulting, and John Reilly
MIT, 2/1999 p. 1

[217] U.S. EPA, Global Warming Impacts, Influencing Factors, Carbon Dioxide Temperature and Crop Yields
Retrieved 3/13/2006 from:
http://yosemite.epa.gov/OAR/globalwarming.nsf/content/ImpactsAgricultureInfluencingFactors.html

[218] Pew Center on Global Climate Change, A review of impacts to U.S. agricultural resources, by Richard M. Adams Oregon State University, Brian H. Hurd Stratus Consulting, and John Reilly MIT, 2/1999

[219] IPCC TAR 2001 10.2.2.1

[220] IPCC TAR 2001 10.2.2.3

[221] Educational Web Adventures, Amazon Interactive, How Rainy is the Forest
Retrieved 3/18/2006 from:
http://www.eduweb.com/rain/rainfall.html

[222] Rainforest Facts
Retrieved 3/18/2006 from:
http://www.rain-tree.com/facts.htm

[223] IPCC TAR 2001 Chapter 9, Executive Summary

[224] Columbia University, CIESIN, reprint of the WHO 1990b. Tropical Diseases 1990. TDR-CTD/HH 90.1. Geneva: World Health Organization.
Retrieved 3/18/2006 from:
http://www.ciesin.org/docs/001-614/001-614.html

[225] Climate Hotmap
Retrieved 2/22/2006 from:
http://www.climatehotmap.org/camerica.html

[226] University of Wisconsin-Madison, 11/16/2005, Third World bears brunt of global warming impacts
Retrieved 2/22/2006 from:
http://www.news.wisc.edu/11878.html

[227] WHO Flooding and Communicable Diseases Fact Sheet
Retrieved 3/18/2006 from:
http://www.who.int/hac/techguidance/ems/flood_cds/en/

[228] WHO Flooding and Communicable Diseases Fact Sheet
Retrieved 3/18/2006 from:
http://www.who.int/hac/techguidance/ems/flood_cds/en/

[229] Wikipedia
Retrieved 3/18/2006 from:
http://en.wikipedia.org/wiki/European_heat_wave_of_2003 and
New Scientist, 10/10/2003, European heat wave causes 35,000 deaths
http://www.newscientist.com/article.ns?id=dn4259

[230] Environmental Health and Safety Online, Global Warming Impacts on Human Health
Retrieved 3/18/2006 from:
http://www.ehso.com/ehshome/globalwarmingimpactshealth.htm

[231] DIS, Northridge Earthquake facts
Retrieved 3/18/2006 from:
http://www.dis-inc.com/northrid.htm

[232] California Earthquake Authority, General Information, original source: California Dept. of Insurance

[233] Insurance Information Institute, Earthquakes: Risk and Insurance Issues, March 2006
Retrieved 3/18/2006 from:
http://www.iii.org/media/hottopics/insurance/earthquake/and also from DIS, Northridge Earthquake facts
http://www.dis-inc.com/northrid.htm

[234] California Earthquake Authority, General Information, original source: California Dept. of Insurance

[235] Hurricane Insurance Information Center, Hurricane FAQs
Retrieved 3/18/2006 from:
http://katrinainformation.org/disaster2/facts/katrina_faq/

[236] Besides a constant barrage of news on this subject, the following is a prime example.
USA Today, After Katrina, insurance tops family's list of tough battles By Larry Copeland
Retrieved 3/18/2006 from:
http://www.usatoday.com/news/nation/2006-02-15-katrina-penrose_x.htm

[237] Seattle Post Intelligencer, Sunday, March 19, 2006 "Reality of climate change hits insurers" by Derrick Z. Jackson

Retrieved 3/18/2006 from:
http://seattlepi.nwsource.com/opinion/263431_jackson19.html

[238] FEMA, October 1991, "Projected Impact of Relative Sea Level Rise on the National Flood Insurance Program"

[239] IPCC TAR 2001 Chapter 8. Executive Summary

[240] Emissions derived from a variety of sources including:
BBC, "Climate change: The big emitters" 7/4/2005
Retrieved 3/19/2006 from:
http://news.bbc.co.uk/1/hi/sci/tech/3143798.stm
Also UNEP from: http://www.grida.no/db/maps/collection/climate6/index.htm
And UNEP from: http://geodata.grid.unep.ch/page.php
And EIA http://www.eia.doe.gov/pub/international/iealf/tableh1co2.xls

[241] Energy Information Administration (EIA), based on 2004
Retrieved 3/19/2006 from:
http://www.eia.doe.gov/emeu/cabs/Usa/Oil.html

[242] Energy Information Administration (EIA), based on 2004
Retrieved 3/19/2006 from:
http://www.eia.doe.gov/emeu/cabs/Usa/Coal.html

[243] Rocky Mountain Institute, Energy, "Is Hydrogen Dangerous"
Retrieved 3/18/2006 from:
http://www.rmi.org/sitepages/pid536.php

[244] Energy Information Administration (EIA)
Retrieved 3/19/2006 from to sources:
http://www.eere.energy.gov/afdc/altfuel/gas_benefits.html and
http://www.eere.energy.gov/afdc/afv/gas_vehicles.html

[245] MSNBC from Associated Press, "Not all hybrids alike with new tax credit" 1/6/2006
Retrieved 3/19/2006 from:
http://www.msnbc.msn.com/id/10692380/from/RL.2/

[246] MSNBC "Boost for natural gas cars: Home fueling" 9/10/2004
Retrieved 3/19/2006 from:
http://www.msnbc.msn.com/id/5960905/

[247] U.S. Dept. of Energy, Alternative Fuels Data Center
Retrieved 3/19/2006 from:
http://www.eere.energy.gov/afdc/progs/search_type.cgi?1/E85_GSLN

[248] U.S. Dept. of Energy, Alternative Fuels Data Center
Retrieved 3/19/2006 from:
http://www.eere.energy.gov/afdc/e85toolkit/environment.html

[249] Ethanol Fuel from Corn Faulted as 'Unsustainable Subsidized Food Burning' referring to studies by Roger Segelken from Cornell University
Retrieved 3/20/2006 from:
http://healthandenergy.com/ethanol.htm and the original press release from Cornell at:
http://www.news.cornell.edu/releases/Aug01/corn-basedethanol.hrs.html

[250] U.S. Dept. of Energy, Alternative Fuels Data Center
Retrieved 3/20/2006 from:
http://www.eere.energy.gov/afdc/altfuel/bio_benefits.html

[251] National Hydropower Association
Retrieved 3/20/2006 from:
http://www.hydro.org/hydrofacts/facts.asp

[252] Nuclear Energy Institute
Retrieved 3/20/2006 from:
http://www.nei.org/index.asp?catnum=2&catid=106

[253] Oakridge National Laboratories,
Retrieved 3/20/2006 from:
http://www.ornl.gov/info/ornlreview/rev26-34/text/colmain.html

[254] Scientific American, February 14, 2005 "Capturing Carbon Dioxide: Efforts to promote an existing technology for limiting power-plant emissions heat up" By Don Monroe
Retrieved 3/20/2006 from:
http://www.sciam.com/article.cfm?articleID=0000492C-072B-120D-872B83414B7F013B

[255] Wikipedia, "Carbon capture and storage"
Retrieved 3/20/2006 from:
http://en.wikipedia.org/wiki/Carbon_Capture_and_Storage

[256] Wikipedia, "Carbon dioxide sink"
Retrieved 3/20/2006 from:
http://en.wikipedia.org/wiki/Carbon_sequestration

[257] Wikipedia, "Carbon dioxide sink"
Retrieved 3/20/2006 from:
http://en.wikipedia.org/wiki/Carbon_sequestration

[258] U.S. EPA LMOP Accomplishments
Retrieved 3/21/2006 from:
http://www.epa.gov/lmop/accomplish.htm

[259] U.S. EPA LMOP Accomplishments
Retrieved 3/21/2006 from:
http://www.epa.gov/lmop/accomplish.htm

[260] All examples of LMOP successes were researched through the EPA's site:
http://www.epa.gov/lmop/proj/prof/index.htm

[261] U.S. EPA "Pay as you Throw"
Retrieved 3/20/2006 from:
http://www.epa.gov/payt/tools/factfin.htm

[262] Wikipedia, Kyoto Protocol
Retrieved 3/21/2006 from:
http://en.wikipedia.org/wiki/Kyoto_protocol

[263] U.S. DOE, EIA, 2003
Retrieved 3/20/2006 from:
http://www.eia.doe.gov/pub/international/iealf/tableh1co2.xls

[264] EUROPA, Climate Change
Retrieved 3/21/2006 from:
http://europa.eu.int/comm/environment/climat/kyoto.htm

[265] The White House, Press Release June 11, 2001, "President Bush Discusses Global Climate Change"
Retrieved 3/21/2006 from:
http://www.whitehouse.gov/news/releases/2001/06/20010611-2.html

[266] Wikipedia, Kyoto Protocol
Retrieved 3/21/2006 from:
http://en.wikipedia.org/wiki/Kyoto_protocol and "China's Kyoto Protocol Hypocrisy" by William R. Hawkins, February 22, 2005 from:
http://www.americaneconomicalert.org/view_art.asp?Prod_ID=1304

[267] The Heritage Foundation, "Why President Bush Is Right to Abandon the Kyoto Protocol" by Charli E. Coon, J.D. Backgrounder #1437 May 11, 2001
Retrieved 3/20/2006 from:
http://www.heritage.org/Research/EnergyandEnvironment/BG1437.cfm

[268] Personal Level steps to reduce greenhouse gas emissions, including the amount of reductions was primarily researched from the U.S. EPA's web site at:
http://yosemite.epa.gov/OAR/globalwarming.nsf/content/ActionsIndividual.html

[269] Environment Canada, Science of Climate Change FAQ Question B.4
Retrieved 1/28/2006 from:
http://www.msc.ec.gc.ca/education/scienceofclimatechange/understanding/FAQ/sections/2_e.html

[270] IPCC TAR 2001 2.3.3
Retrieved 3/23/2006 from:
http://www.grida.no/climate/ipcc_tar/wg1/070.htm

[271] NOAA, NOAA ATTRIBUTES RECENT INCREASE IN HURRICANE ACTIVITY TO NATURALLY OCCURRING MULTI-DECADAL CLIMATE VARIABILITY Nov. 29, 2005
Retrieved 3/24/2006 from:
http://www.magazine.noaa.gov/stories/mag184.htm

[272] NOAA
Retrieved 3/24/2006 from:
http://www.cpc.noaa.gov/products/outlooks/figure2.gif and
http://www.cpc.noaa.gov/products/outlooks/hurricane.shtml

[273] University of Leicester, "Greenhouse Theory Challenged by Biggest Stone"
Retrieved 3/23/2006 from:
http://www2.le.ac.uk/ebulletin/news/press-releases/2000-2009/2006/03/nparticle-bxh-khs-ykd

[274] NASA Goddard Institute for Space Studies, GISS Surface Temperature Analysis
Retrieved 3/23/2006 from:
http://data.giss.nasa.gov/gistemp/graphs/

[275] Softpedia, "A New Explanation of Global Warming" by Vlad Tarko, 3/14/2006
Retrieved 3/23/2006 from:
http://news.softpedia.com/news/A-New-Explanation-of-Global-Warming-19650.shtml

[276] Nature 438, 655-657 (1 December 2005) | doi:10.1038/nature04385 "Slowing of the Atlantic meridional overturning circulation at 25° N" Harry L. Bryden

[277] Referring to: Science 17 June 2005: Vol. 308. no. 5729, pp. 1772-1774 DOI: 10.1126/science.1109477 "Dilution of the Northern North Atlantic Ocean in Recent Decades" Ruth Curry and Cecilie Mauritzen
From RealClimate: http://www.realclimate.org/index.php?p=191

[278] Science 13 August 2004: Vol. 305. no. 5686, pp. 953-954 DOI: 10.1126/science.1100085 "Enhanced: Already the Day After Tomorrow?" Bogi Hansen et al

978-0-595-40622-7
0-595-40622-X